The Showman Shepherd

The Showman Shepherd

DAVID TURNER

Farming Press

First published 1990, reprinted 1992

British Library Cataloguing in Publication Data
Turner, David
The showman shepherd.
1. Livestock.   Sheep.   Showing
I. Title
636.318
ISBN 0-85236-204-8

Published by Farming Press Books
Wharfedale Road, Ipswich IP1 4LG
United Kingdom

Distributed in North America
by Diamond Farm Enterprises,
Box 537, Alexandria Bay, NY 13607, USA

Cover design by Mark Beesley
(front) *Mrs Rosemary Parkes and Andrew Read,
representing Lincoln Longwools in the Interbreed Championship,
confer with a steward*
(back) *A strong line-up of Hampshire Downs*

Typeset by Galleon Photosetting, Ipswich
Printed and bound in Great Britain by Biddles Ltd,
Guildford and King's Lynn

Contents

Colour plates appear following pages 22 and 54

Acknowledgements

I should like to express my gratitude to the following people, who have been of great assistance to me in the preparation of this book. Most did not know me from Adam, but all were kindness itself and most knowledgeable. What is more, they were prepared to share their knowledge. What seemed clear was, they all love sheep.

Jim Dufosee	Joan Hayes
Eric Halsall	Graham Rowles Nicholson
Jane Paynter	Mrs J. S. Brigg
George Hughes	Billy Bland
Hugh Clarke	J. F. Richardson
John Randall	Mrs A. M. Burnham
Sarah Randall	Neil Fairbairn of Youngs
Wally Powner	Chris Lewis
Ron Goodwin	Lesley Stubbings
Brian Holgate	John Terry

Most important of all, special thanks to Ena, my secretary.

To Judith, the best wife a shepherd could ever wish for

Foreword

I first became a shepherd through the Rare Breeds Survival Trust. I began by wanting to preserve a rare breed and solve a grass problem at the same time. This led me to look closely at sheep and eventually brought me to choose the Southdown as the breed for me. Time has merely reinforced my love for this breed, and an added bonus has been the extent to which preparing an animal for shows and showing is an important feature of the breed.

Some sheep are highly sought after and numerous, others are rare but also easy to sell. My breed, the Southdown, sells best when maximum effort is made in the presentation of the animal. Showing is therefore absolutely vital if one wishes to gain maximum enjoyment of the breed and maximum profit at the same time.

As a general practitioner, used to learning from print, I looked in vain for an instruction manual on showing sheep. None was to be found, and I therefore had to learn the hard way, by trial and error. I'm sure I must have raised a few eyebrows to start with; I probably still do.

As a consequence, I have resolved to commit to paper all that I have learned so far, just as Val Stephenson took the trouble to pass on her findings in the *Shepherd's Calendar* when the freshness of the early trials and tribulations was still in her mind, and I feel that even the modest amount I have up until now learned about showing sheep is worth passing on to others, especially the novice teetering on the brink.

I hope I have done my research well enough not to have committed too many sins on the following pages. Many of the ideas you will find have been contributed by some of the most eminent and successful sheep people in the land. I have no doubt that some of my own ideas are contentious and that the work is by no means comprehensive. If, however, it provokes interest and discussion, encourages further research and, most importantly, helps fellow novices, I shall be well pleased.

DAVID TURNER
November 1989

Why Show at All?

There's no business like show business! This is true, even in the field of agriculture (no pun intended). The fact that you are reading this must mean that you will have attended some county shows, perhaps even the Royal Show, and will have enjoyed not only the exhibits but, I have no doubt, the atmosphere too. I promise you, until you have been fully involved in a show as an exhibitor, you cannot yet have experienced the full thrill. The carnival-like atmosphere at Stoneleigh the night before the first day of the Royal Show must be felt to be believed. The camaraderie within the animal lines, among the competitors, is just wonderful. Everyone feels both a sense of excitement, and that they somehow belong. People are so willing to lend a hand, or just provide information, especially to the novice. The reason is obvious; the more competition and the better the class of competitor, the greater the glory in winning. I prefer to come third in hot company than win in a pushover. Of course, you may hear lots of amusing tales about 'dirty tricks' and sabotage, but I have yet to see any. The accent is always on friendship and support. In particular, the newcomer is made most welcome, and when he or she asks for advice, this is flattering indeed, and it is readily available.

If I am beginning to sound like Pollyanna, suggesting that everything is 'just wonderful', let it merely be a sign of my enthusiasm. All breed societies are like families; they have their ups and down, and some have their fair share of, dare I say it, black sheep. I have even heard backbiting!

Generally though, the warmth and kindness shown to me over the last few years by the vast majority of showing shepherds has been the abiding memory, and writing this, I am looking forward with some excitement to the next show season. I have met so many super people on the circuit that I look forward to meeting again.

The shop window   If you breed sheep you have to sell them. What better place can there be to show off your wares than the

1

showground? Farmers may be a conservative lot, but they are always keen to see available stock, and when they stop and take an interest in my Southdown rams, I feel that I am dealing with a possible prospective customer, supporting the breed and keeping it in the mind's eye.

The public It may be easy to perceive the public as ignorant, sentimental, interfering and fickle hypocrites. They want to eat cheap lamb from abroad, never minding the quality; they moan about the use of chemicals and 'ooh' and 'aah' about the baby lambkins all at the same time. It is our duty to educate them, and shows are the ideal place to promote the concept of the better-quality meat, home-grown by thoughtful, caring farmers.

Continuing education It is very easy to live in a rut, going on year after year doing the same old things, without ever being critical of whether it is still the best way. Take your stock and yourself, and compete in a show where you will have the opportunity to test some of your theories against those of others. The show is a real learning experience from which even the wisest among us can draw benefit. There are books and courses that we can attend, and even hanging about the sheep-lines you may pick up snippets of useful information.

None the less, you only really get to learn once you are there competing and committed, and so I hope to get the 'tempted-but-cautious' to take their first step. I promise you, the thrill of being in the ring under the judge's eye, and even better, gaining a rosette, is something to experience. Later on, to win a championship rosette, and to stand chatting with the judge, is very sweet indeed.

WIN, LOSE OR DRAW

A wise stockman told me some advice given to him about the competitive side of showing: 'If you win, say nothing. If you lose, say less.' Judges are not perfect, and sometimes you will disagree with their judgement. However, like the umpire, they are always right, and besides, you have to set yourself out to appeal to them, if you can. They have their own tastes, which may not always coincide with yours. Be modest in victory, realising that it is a modest acknowledgement of one man's view. Coming last does not mean an animal is bad, only that you must work to produce animals of a higher standard that would appeal to a wider range of judges. It may be easy to sell a prize-winning animal at a show, though curiously it is often easier to sell a slightly less successful animal. Even if you have not caught the judge's eye this time, your stock has at least been under the gaze of the public. I like to spend a fair amount of the time around my stock and to wear a name tag, so that the general public may ask questions and become more aware of my animals. After all, farmers who stop to chat may be potential customers.

I assume that you are going to breed pedigree sheep, which will need to be presented in their best light against others of the same or similar breed, age or sex. There are classes in shows of crossbred or 'end product' sheep which are normally shown in their working clothes, just tidy off the fields if necessary.

Generally the fifty-seven varieties of sheep seen in shows are kept pure because of certain specific breed characteristics, and showing attempts to demonstrate these to the full. The Southdown is known for its very fine short wool and the meatiness of the carcass. All *my* efforts, therefore, are directed towards proving that my Southdowns have these characteristics at the highest level.

WHERE TO SHOW

You can show at several levels, beginning at the small local shows with little distance to travel, small entry fees, an informal attitude and they only last a day (very often Sunday). These are great fun and offer a good place to start, but there are problems. Classes will be small, and they are often made up of several different breeds, possibly lumped together under an 'Other Pure Breeds' label, which comes after the Suffolks and Jacobs, and occasionally after other specific classes with local support have been judged. They will often be held out of doors, so you could be drenched, and the judges tend to be slightly less expert. Organisation can be chaotic, and unless you are pushy enough to keep pestering the stewards, you might miss your class, or you may merely become flustered, feel insecure and leave the showground angry, determined not to bother again.

In comparison, the larger shows that may *seem* daunting, are often under cover, beautifully organised and stewarded, with excellent judges and facilities. However, the entry fees are higher, and the competition stronger, but here you will learn the very best way to show sheep by being in the company of the top sheep people. You may have to travel a good distance, and the show may last over several days. I do a mixture of both these types of show to gain the advantages of both. I am happy to muddle along in the 'Little Pickle in the Wold' show, accepting it in the spirit in which it is attempted, and the following week I will be pleased to be at the Royal Show. I take both types seriously, and I try to win one hundred per cent at both.

WHEN TO SHOW

The show season begins in the spring when the weather is usually wet and cold, and it goes right through to autumn. At the back of this book you will find a comprehensive list of shows and dates. Your own breed society will patronise certain shows more than others. Some shows are also linked to official sales of the breed. The Rare Breeds Survival Show and Sale is definitely one of the high spots of my showing year. Many friends meet there, and there are so many breeds that it never fails to excite me. Stoneleigh is a marvellous show venue, and with the show on the Friday and the sale on the Saturday, it really constitutes a great thrill.

How to Start

Many people would want to wait until they have bred their own animals before they show, but there is no reason why, so long as you declare the breeding, you cannot show sheep bred by other people and brought on to show peak by yourself. This will normally mean raising a bought-in lamb to adulthood, so you will have good reason to take some of the credit. What would be objectionable would be to buy the Royal Show champion ram and then show him at a local show!

I have less time for folk who don't show unless they have wonderful stock. I believe in showing my best every year; some years they are better than others. Of course, all animals being shown must be healthy, sound in limb, true to type and free from obvious defects, e.g. undershot/overshot jaws. When you show a pair of animals, they ought to be as alike as two peas in a pod, but rarely are. I pick the most alike I can get, which may not necessarily include my best individual, but I would probably not show my two worst lambs simply because they were identical, nor would I refuse to show at all.

My advice would be to get into showing as soon as you can. The sooner you start, the sooner you learn. Don't be put off too early by some of the high-flown techniques, such as trimming. Before I learned to trim, and being unaware that people sheared on the first day of January for the Royal Show in July, I showed a ram with a mere five weeks of wool on him. He was practically naked, but still he won a championship. Incidentally, I took the same ram back to the same show the following year, again naked, and again he won the championship!

Shearing and housing It is customary to show sheep with seven months of wool on them at the Royal Show, particularly the shortwool breeds. This effectively means that the sheep need to be sheared on January 1st, and our breed society stipulates that they may not be sheared before this date. Naturally, if you are going to be shearing on January 1st, the stock will need to be kept indoors for about eight weeks afterwards. By then there is enough wool to keep them warm when turning them out. In any case, it makes some sense to keep them in at this time of year, because the grass is largely valueless as a fodder, and they are likely to be subject to all sorts of foot problems.

Some farmers believe that sheep are more prone to foot-rot if housed on straw, and they footbath them every three weeks. I do not do this, but then I probably use more straw than necessary to ensure a dry bed.

The great advantage to me of housing my sheep is that I can keep a very close eye (and a frequent hand) on them, and so closely monitor their progress. They become more docile and familiar with handling, and this will stand in good stead during the show period.

Housing

Some people house the sheep from January 1st shearing, and never let them out again till after the show season, their object being to increase size and accustom the sheep to being penned, handled and under cover. Other people feel strongly that this is artificial, and prefer to take their show sheep out of the fields without preferential treatment, tidying them up within a few days, and getting them used to the new way of life. This second school content themselves that they are in business to raise fit sheep to do a job, and that the show circuit is merely an extension of what they normally do to sell a sheep. They scorn the intensive preparation some others use, and accuse those who mollycoddle their animals of producing a sheep that will be unfit for its purpose. There is no doubt that stuffing corn into a ram through the winter and spring, and lugging it around all the shows, could result in a level of unfitness that renders it incapable of serving any ewe, let alone forty or fifty.

Surely, the answer is 'horses for courses'. The lowland sheep, bred for an easier climate and taking naturally to folding, is eminently suited to the full housing and feeding regime, particularly if shorn bare on January 1st. The hardy northern hill sheep, like the Herdwick, Lonk and Derbyshire Gritstone, serve their purpose on the hills and are best shown in their working clothes straight from their usual environment. Some breeds fall into the middle ground, where a little more effort and expense makes a good deal of difference, without breaking the pocket, taking hours of valuable farming time and ruining the sheep for its purpose.

Within reason, any sort of dry building will do to house sheep, but there are some factors to be taken into consideration.

Sheep are designed for outdoor living, and tend to be subject to stress when housed, at least initially. The stress is less if they are sheared, and it is salutary to look at two housed flocks, one sheared and one not, as I have done recently, and notice the much lower rate of breathing in the sheared flock.

A building is to keep rain and snow off, and a 'lean-to' will be

7

better than a stable because, if it faces away from the prevailing wind, it will be cool and airy, but not draughty.

Sheep need between 10 and 15 square feet of space each, and about 18 inches of space at the trough at feeding time. They will need to be well bedded in dry straw, which they will also eat to some extent when fresh.

It goes without saying that fencing must be secure, and make sure that as the level of straw bedding rises, you raise the barriers too, particularly if you are housing both sexes adjacent to each other. I have been most impressed by Netlon Tensar, a heavy-duty black perforated plastic cladding, which is both stockproof and capable of reducing any gale to about 7 m.p.h. maximum. Recent weather has borne out the manufacturer's claim that it is also snowproof. I have made up light rectangles of timber frame, and tacked on the cladding to fill in gaps into which snow would be driven, with great effect.

I feed hay in improvised racks, made by angling one hurdle against another used as a barrier. They are held close together at the bottom and wider apart at the top with the ever popular binder twine. Hay is fed pretty well ad lib, though this does produce choosy feeding and some wastage of the tough bits. On the other hand, these tough bits end up as extra litter on the floor, and that's pretty well all they are fit for in any case. Hay-making was awfully late in 1988 (Royal Show week), and so a bit over-blown and not too nourishing. Often the barley straw tickles the palate more—swings and roundabouts.

Walk-through feeders I have fed cereal in bowls and buckets, and it works for one or two animals, but sheep always think someone else is getting more, and they tend to fight and spill food everywhere. This winter I set to and built some walk-through feeders out of pallets, and I have been delighted with them. The best size is at least 3 ft by 3 ft, and better still 4 ft by 4 ft. Try to get several pairs of identical ones with fairly open slats already. Remove the middle and one of the end blocks, then, standing the pallets up on the remaining block, link this together facing inwards. Use the pieces you have removed as planking; if you damage these, and I often do, you still need to prise off a plank towards the bottom to allow the sheep to get their heads through, and then these can form the floor (*see diagram*).

The walkway needs to be at least hip-wide; mine are about 18 inches between uprights. I use the blocks that have been prised off to produce extra support for the floor planking, and strengthen at the bottom with cross-pieces from the planks

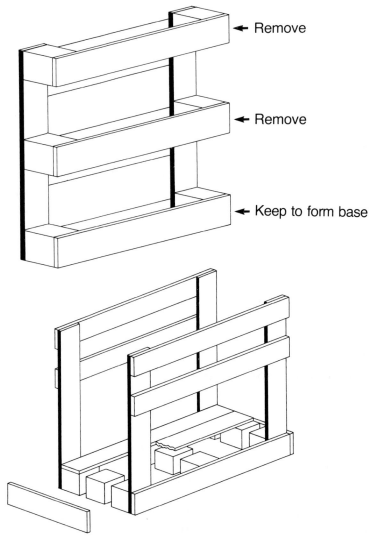

← Remove

← Remove

← Keep to form base

Walk-through feeder construction

removed earlier. The great art in this is not the amazingly easy carpentry, but getting, say, six good-quality *matching* pallets to make three identical feeders, and design the feeder as you go, depending on the measurements of the pallets.

Of course, you could have one-sided feeders using a single pallet, with minimal extra ingenuity.

Lighting   This is essential because, at the time of year when housing is important, there is little light and, if like me, you are not a full-time farmer, or perhaps even if you are, then many jobs will have to be done after dark. Feeding is probably best done, as with milking cows, every twelve hours to get the best out of the feed, and you know how animals love a regular feeding routine.

I have tried all sorts of temporary lighting ideas using Calor Gas, Tilley lamps, and battery-powered strip lights which are very useful in caravans or helpful as an emergency at the road-side, working off your car battery. Let me say that, regardless of the expense, for sanity's sake push the boat out and go for mains electricity. To economise I use the two-core orange cable you run garden appliances off, because it is cheap, easily seen and safe enough to run a few light bulbs off without heating up, especially in winter. For extra safety I have it plugged into an earth leakage circuit breaker, which anyone who uses garden appliances ought to use anyway.

Thank goodness it is really easy to do your own wiring for light. Bring in your source of supply to a junction box screwed to a beam high up. From there you merely need to take a supply to a switch, then to the light (*see diagram*). All the requisites cost a very few pounds, and the end result is well worthwhile.

To shear or not to shear   I recently visited a local sheep farmer friend to see 300 housed ewes due to lamb in seven weeks (March) who had been sheared five weeks previously. I must say that they looked well relaxed with a low respiratory rate, and there was no sign of wool slip anywhere. In previous years wool slip had been a problem, probably related to a nasty cold spell just after shearing. It is believed that the extra stress of the cold increases the body steroids, and affects the new coat badly,

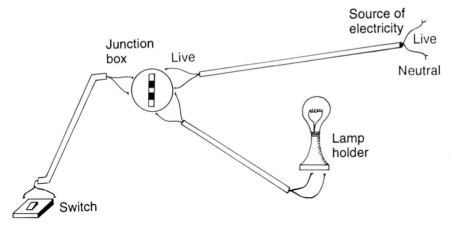

It is vital that the switch be placed on the live
side, so that the bulb is actually isolated from
the mains supply when the switch is off. If you
are in any doubt about this, do go for safety and
invest in an electrician.

Wiring diagram showing connection to junction box

producing slip. This would be got round by rugging up show
stock. Incidentally, as my farmer friend has taken to shearing
his in-housed ewes, he can stock more densely—a mere 10 sq
ft per ewe—less trough room is needed, and the wool quality
has risen by four points in comparison with spring shearing. Of
all the locals who have converted to winter shearing or housing,
none have reverted to their old ways.

Housing sheep, however, is not without its problems. Sheep
are naturally inclined to be outdoors and, particularly in full
fleece, they are protected against the weather. Indoors they are
much more prone to disease; you will use a lot of straw and a
great deal more feed, and you will have to keep a very close eye
on them.

Feeding

On the subject of straw, look at the figures for the food value of straw, and you'll understand why, if the straw is clean and fresh, the sheep will eat a good bit of it. It has been shown that letting the sheep pick over their litter works well, as they select the best bits. No additive or chopping up does better.

About three weeks before the date that you expect to house your sheep, you should begin to feed them some molassed sugar-beet shreds. Many show people incorporate this into the diet of the animals they wish to bring on well, particularly if they are to be housed. The molasses helps the rumen, tending to stabilise it in the face of a rather odd diet of hay, straw and cereals. You may know that the action of the rumen is dependent on the lacto-bacillus, which actually digests the cellulose in grass, releasing free fatty acids, mainly propionic and acetic acid, which can then be converted into energy or stored as food. Anything that can keep these lacto-bacilli happy, and molasses certainly does, will help utilise the basic forage. Not only that, but the pH or acidity of the rumen stays low with all these free fatty acids about, and thus helps prevent gastro-intestinal diseases, including scouring. The upshot of this is that you can increase the cereal input so that the growth is maximal, without getting into too much trouble.

Molassed sugar-beet shreds contain 11 per cent protein, so it is sensible to mix this with 16 or 18 per cent cereal feed. I use about 50 per cent, though the British Sugar Corporation recommend 25 to 40 per cent for use in the in-lamb ewe. One *very important* precaution with sugar-beet shreds must be taken, and it is this: *always soak the sugar-beet* at least overnight (about one gallon slowly poured over three kilograms of dry shreds works well). Manufacturers do make it clear on the bags that for horses it must be soaked, and several friends have got away with feeding it dry to sheep and cows who do salivate, but feeding it dry has cost me money.

A cautionary tale I had been trough-feeding dry sugar-beet

Pen gate complete with feed trough, hay rack and bucket held above dunging height by a simple wire holder

This feed trough poses no problems for the DIY handymen to build

shreds for several weeks just prior to housing when one evening I noticed odd behaviour, first in a young ram and then a ewe. They shook their heads, pawed the air and generally leapt about, but made no sound. Their breathing was unimpaired. The ram soon cleared his gullet spontaneously, but the ewe began to froth at the mouth, and eventually shook out or sneezed out long strings of disgusting white mucousy catarrh. After half an hour or so, despite my attempts to massage down her neck, she was clearly bloating up on the left side. By one and a half hours I was glad to be in my vet's operating room, where an antispasmodic injection plus rather brutal use of a large rubber tube down the throat eventually cleared the blockage, releasing great volumes of gas. I have never made the same mistake again.

The same problem can easily occur with hungry sheep going for sliced apple, as a fellow competitor next to me recently discovered at the Royal Show. On this occasion, the bloating began after a mere thirty minutes, but the blockage thankfully cleared just as the vets arrived with their tubes. They sportingly charged no fee! Strangely enough, whole apples on the ground in an orchard do not seem to cause the same problem, perhaps because the sheep are never so ravenous as a penned sheep, and they nibble away at their leisure.

Most people's criteria for judging feeding are a little vague, jokingly referred to by Leslie Stubbings of ADAS as 'all muck and magic'! Some simply rely on:

- *Feeding only to appetite.*
  Some people just stuff it in up to the maximum, but this is probably wasteful. If the sheep don't clear it up, take it away and reduce the subsequent feeds.
- *Allowing hay ad lib in a rack, never a net.*
  If you feed lambs using a net, they tend to climb up it, get their legs caught and then kick even to the point of fracturing a limb. Adult sheep are less likely to injure themselves this way, and besides, their limbs are stronger. By far the better bet is to use a rack and, what is more, to place it low down to prevent seeds and chaff falling into neck wool and eyes. You will occasionally see a lamb with a watery eye, and this could well be due to a seed being lodged under the eyelid.

  A very good piece of equipment was introduced to me by John Christmas of the Kent halfbreeds. He has a sheeted hay rack which hangs low down on a hurdle, and the sheep feed through the hurdle bars (*see diagram*).
- *Always allowing plenty of clean water.*
  Sheep frequently dung in their water, so I check regularly, and I change it at least daily.

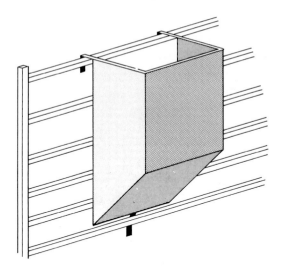

Hurdle-mounted hayrack

High automatic drinkers would get round this problem, but they are expensive. I like to use ex-glucose lick buckets because they are white, and thus show whether or not the water is dirty. Feed as often as you can, by which I mean, split up what you have decided will be the daily concentrate intake into as many different feed times as is practical for you to allow. Keep feeding utensils scrupulously clean. I do not think it possible to skimp with feeding, either on the quality or the quantity, and although it may seem as if you are pumping a lot of money into an animal, there is no doubt that if this animal is to be your showcase, then your money will be well spent.

While the above rules follow for any sheep, there are other guidelines that may help you plan your show stock diet that are more scientific.

It has been known that breeders have fed additives such as arsenic or vitamin B_{12} to enhance the appetite and make sheep fatter quickly before shows. I believe this to be most foolhardy and against the best principles of animal care and service to the buyer and indicate a poor attitude to the show and its purpose.

There seem to be as many ways of feeding sheep as there are shepherds. Farmers with acres of grass and some arable have no problems with hay, straw and cereals, which can all be found on the farm, and under personal supervision with regard to quality control. Other crops used are turnips, various brassicas including rape, and a sort of Chinese cabbage called typhon. Ideally these crops are daily strip-grazed so that one small area is well polished off and the sheep are moved onto a fresh strip the next day. Some go to the extreme of moving the fence twice a day.

It is clearly a great advantage to turn out your lambs at weaning onto a crop of, say, turnips with up to 1 lb of concentrates extra to the rams. Those of us who only have pasture, however, can take some consolation from the fact that the sheep's natural diet is grass and the herbs found wild in a well-managed field. Just think for a minute about human beings. We can deduce from our dentition that we have evolved as omnivores, but we know that eating large quantities of animal fats shortens our lives, and that vegetarians come to no harm. Our bodies are capable of healthy existence so long as there is just enough of everything, i.e. protein, fat, carbohydrate, vitamins and minerals, and are sophisticated enough to be able to break down the foodstuffs, if needed, into simple units, and reforming those into the more complicated tissues.

If you have been confused about degradable and undegradable protein, this is what it is all about. The vegetarian's proteins are all degradable and are, at small effort of the liver, broken down into smaller units, and rebuilt to be the proteins of muscle, bone, skin, etc. If he eats meat, larger units can be utilised. As I have suggested, however, it does no harm to be a vegetarian, and conversely, it does no harm for a sheep to eat fish meal, though not by any stretch of the imagination could you envisage a sheep catching a fish, let alone eating it!

In fact, research has demonstrated how valuable fish meal is to sheep, particularly in the lactating phase, where the quality of milk is markedly improved. Good-quality lamb feed ought to contain at least 2½ per cent fish protein.

VALUE OF GRASS

It does not follow that you *have* to feed sugar-beet, rape, typhon, turnips or even cereals, but it helps to have a mixed diet, particularly at the time of year when the grass contains less protein. It is said that there are but six weeks a year when the grass is any good and, of course, this is when we try to conserve for hay, and make it before the effort of the plant has already gone to making the seed. This is like many things one hears in agriculture, somewhat of an exaggeration. I have even been told by a loquacious butcher that hay must be cut at 11 a.m., because then it has the highest protein of the day! Where does he think the protein goes to? Where does he think it comes from?

The reality is that at all times of the year grass is capable of providing fibre, water, vitamins, minerals, some protein and energy. It contains about 20 per cent dry matter only as compared to the 85 per cent of hay. While its carbohydrate content is reduced in winter to about two thirds of its best, its protein level falls to nearly a third (but doesn't reduce to zero). This means that to get the same protein the sheep needs to eat three times as much grass in winter, which takes longer, particularly if the sward is very low, and consequently leaves very little time for cudding.

Recent research has shown that you can base your supplementary feeding on actual grass height when assessing how to feed milking ewes after delivery.

Grass height (cm)	*Concentrates/kg/ewe/day*	
	Twins	*Singles*
3 or less	1	0.3
3 to 4	0.5	0
4 or more	0	0

These figures were produced in relation to lamb weight-gain, and suggest an optimum. It was also suggested that during lactation the protein percentage of the feed be increased 4 per cent over what had been fed before lambing (up to 20 per cent), and this could be achieved by using a 34 per cent protein mixer. The extra protein apparently enables the milk to come 'off her back'.

Of course, underfeeding leading up to lambing not only brings the risk of twin lamb disease, but also it has been shown that it means:

  10—20 per cent more energy is needed in lactation because of low body reserves
  18—20 per cent decrease in udder size
  10—30 per cent fall in milk yield

To sum up then, the feeding of the show animal begins at flushing and conception, continues through pregnancy to culminate in a well-fleshed strong lamb born to a well-fed milking ewe who then, for 12—16 weeks, is fed according to grass height a higher protein concentrate to maximum lamb growth. During this time some cereal creep feed is introduced to the lamb, starting at about ten days. It has been noticed that cooked feeds, such as a coarse mix, are taken most avidly, and many people use them. There are disadvantages—most importantly, care must be taken with coarse ration because, unless the maker has specified that it is suitable for lambs, it will probably contain minerals at the high levels only suitable for adult sheep. The sad outcome of feeding this ration may be that perhaps your most greedy, hungry ram lamb—your pride and joy—develops kidney stones and dies. Feeding coarse ration 50 per cent with cereals is safer, so is checking mineral levels with the manufacturers. Secondly, they cost up to £40 per tonne above ordinary lamb creep feeds, and lambs soon learn to pick over the mix, eating the bits they like best. Sometimes they turn up their noses late in life at plain cake, but not for long!

On the subject of cereals, I do not want anyone to think that I am necessarily referring to proprietary cereal nuts. While a 17 per cent lamb creep feed may do very well for the rapidly growing lamb being shown at 20 to 24 weeks of age, this ought to be a 'start to finish' preparation. Such a high level of protein is not needed for the shearling, and given that such feeds contain about 2½ per cent fish protein, they will tend to lead to a rather flabby soft fat deposited on the back. What most judges seem to be looking for is a hard feel that is made up of hard white fat deposited over the eye muscles. It seems to be true that to produce this kind of hard fat you must feed a starchy feed, preferably a wholegrain. The preferences are whole barley followed by wheat, and oats coming a relatively poor third. The ram will get through 1½ kilos a day of such a feed, made up of:

> Six parts by volume whole barley
> Three parts by volume molassed sugar-beet shreds
> A handful of linseed cake, or less

It is difficult to know for sure why linseed cake is such a good feed. What seems quite certain is that it is very palatable, and probably makes food more interesting. It undoubtedly provides 56 per cent protein and 7 per cent oil, and has been traditionally used by people showing cows and horses, who believe that it imparts more oil to the coat. This clearly is not of any relevance with sheep, but I have noticed that the use of a small proportion of linseed cake brings on the condition of sheep more rapidly than when it is left out. It seems to be agreed that there is no place in a feed at this stage for peas and beans, though there are some people who use them and get good results.

I cannot stress too strongly that whatever you feed, it must be very gradually increased over at least a week, starting with perhaps a third of a kilo every day. There must always be fresh water in abundance, and hay and/or barley straw. Keep a very close eye on the dung, to make sure that it still pellets well. You will occasionally find an animal that will not tolerate hard feed at this sort of level, and will scour readily. It is sometimes difficult to hand-feed a chosen animal away from its fellows, particularly, for instance, if you are trying to rear a pair of matched ewe lambs and one seems to be able to eat more than the other. I believe the answer is to judge the feeding according to the slowest eater, and then try and watch feeding to see whether one is getting more than the other. If this is the case, you must feed smaller feeds more frequently, so that the difference between the shy feeder and the glutton becomes less. Where a lamb is scouring on these sorts of feeds, it may be sensible to reworm, and I have often been surprised by how rapidly after worming a lamb may become re-infested.

I have also found it extremely helpful at the same time to both withhold cereals for a day or so, and to use a probiotic which has the dual effect of increasing appetite for hay and straw, and of enhancing the levels of micro-organisms in the rumen, by effectively feeding them and maintaining the environment in the rumen that they like best. I would also use these probiotics at any time I use an antibiotic, because the antibiotic, while it will kill the bacteria you do not want, will also kill the bacteria you do want.

Above all, the sheep must be regularly handled to ensure that they are moving towards the sort of condition you want them to be in at the show. While lowland breeds tend to be shown at above grade 3½−4, some of the lighter-bodied animals rarely get above 3½.

WATER PROBLEMS

Some people have discovered that offering water ad lib to sheep in show pens can be a mixed blessing and ration it severely or even deprive their sheep completely. They point to cases of severe bloat incurred by lambs occasionally swigging down whole buckets of water and having to be resuscitated smartly by the judicious application of a rubber tube down the throat. Just such a problem happened to Jimmy Wilson at the Royal just forty-five minutes before going into the ring with his Highland Show Champion ram lamb. Urgent and successful first-aid allowed him to compete, win his class, then the male championship and finally, as in all good fairy tales, the supreme interbreed championship!

Some people's anxiety about water stems from the idea that merely a change of water from that which the sheep are used to can make them scour. All I can say is that a change of water doesn't make *me* scour and my suspicion would fall on stress in a neurotic or poorly acclimatised sheep or even, more likely, inadequate hygiene with buckets. In lambs, failure to worm might be the cause. Once you have scouring, opinions differ about treatment. Some favour 'stopping' medicines but I have found probiotics the most successful and I like the idea that they stabilise the rumen by encouraging the 'good-guy' bugs over the pathogenic coliforms such as *E. coli*.

If extra penning is available, it would seem a good idea

to isolate the scouring animal so that not only will the spread of infection be prevented, but neither will other sheep become plastered. Here is where the pen sheets really pay their way, as on the occasion when a Jacob ewe next door to me flung dung everywhere. Washing it off the sheet was unpleasant enough, but imagine trying to wash it off a trimmed ram just prior to judging!

Don't imagine that the anti-water brigade just allow their stock to dehydrate, shrivel up and die. They provide their water in the form of some succulent feed such as cabbages which have the added advantage that they are good for the digestion and tend to be beneficial for the teeth, offsetting the tendency for hay to pull the first two teeth out of line giving an overshot appearance. This last can be quite a problem for some sheep on hay; a ram perfectly sound in the mouth at the start of the season can be thrown out of the ring at the end. Constant vigilance will allow you to change the feeding regime and, for example, within about ten days of normal grazing, the teeth will be back where they belong.

At some of the larger and longer shows 'greenmeat' (freshly cut grass) is supplied daily, usually in the evening. It is a good idea to be around when it arrives because those in the know snap it up pretty swiftly and if supplies are meagre you will miss out. As with the non-waterers there is also a hard core of those who would not risk feeding greenmeat to their stock for fear of digestive upsets. I personally have encountered no problem but always feed it cautiously, at least on the first evening. The sheep devour it very greedily which ought to tell us to be careful but also suggests to me that such feedstuff has been lacking in the diet previously.

At some shows, the greenmeat is provided only on the first day and in excess which means that it sits there festering away particularly if it's hot. On day one it's fine. By day three it has fermented pretty well, smells very fruity and looks a lovely khaki colour. Beware! This stuff is really dangerous. Sheep adore it and I have seen them appear to be rolling drunk afterwards. I wonder if they suffer from hangovers?

Somehow, and perhaps I'm just talking about Sod's Law for shepherds, however careful you have been, stock living on straw develop foot-rot and on arrival at the showground are found to

Jimmy Wilson encouraging his Highland show champion ram lamb into the 'hospital' for emergency treatment after it had swigged down a whole bucket of water and bloated just forty-five minutes before judging. See colour picture for romantic happy ending

be hopping lame. This happened to me at the South of England Show where, the night before judging, one of a matched pair of shearling ewes went lame with scald. I was lucky to have noticed it really but then I was taking them for their evening trot, so perhaps I deserved the nudge of the gods? I happened to be penned next to the pig washer where there was a boiler with a constant supply of hot water. I placed the ewe's head in my trimming stand and placed both her hind feet in a bucket of warm zinc sulphate for a full half-hour. I repeated the dose early next morning, prayed a good deal and was rewarded with the breed championship! For years I had carted zinc sulphate around and never needed it. Purple sprays for foot-rot are also good, particularly the oxytetracycline ones from the vet, but you don't want purple dye everywhere when showing.

All lambs are cute but some are cuter. Mrs Prentice's Greyfaced Dartmoor ewe lamb is about the cutest and is suited to being shown just tidy without elaborate preparation

...tty Garbutt ...monstrates the ...ry that is the ...icester Longwool ...ece

Herdwick shearling rams showing the rich wool colour of the juveniles

The Supreme Interbreed Champions, Royal Show 1989. Mr Wilson's Suffolk ram lamb with Mr Timms' shearling ewe. Behind that happy smile, the memory of some anxious moments?

Chris Lewis displays his fine Texel lamb

Mrs Long with a pair of Norfolk Horns. In the nineteenth century, these were crossed with the Southdown to create a new breed, the Suffolk. And they said they would never catch on!

A beautiful Kerry Hill ram earned Mrs Rosemary Beecher the silverware at the East of England Show 1989. She judged Jacobs at the same event

Mrs Johnston, a much respected judge from Dumfries, strides with a purposeful eye among loose Vendeens. The first year that this breed had classes at the Royal was in 1989, and these were just 'Males, all ages' and 'Females, all ages'. What a daunting task!

Champion Jacobs. Mrs Blacknell's three-shear tup with Mrs Sleightholme's ewe

Not the high-jump event but the usual fun and games witnessed at the judging of the trios of cross-bred or commercial lambs. Points may be awarded for the number of complete somersaults

Gordon Brackenbury handling Mr G. Rowles Nicholson's amazing Romney ram, three-times Royal Show champion, carrying at least 26 lb of wool. The boss himself has the ewe, a typical example of the Limestone Romney.

Devon and Cornwall Longwools. Said to be able to produce more wool per sheep than any other British breed, averaging 20 lb a fleece and going up to an incredible 40 lb

These are the Border Leicesters that won the Longwool section of the Interbreed at the Royal '89. (Left) Alison Smith with her ram. Libby Jones (right) had a wonderful year with this ewe, winning wherever they went into the ring

Geoff and Jackie Riby with their superb Oxford Downs. In 1989, for the second year running they won all the classes and therefore ended up with Supreme and Reserve at the Royal. What consistency!

Grading and Finishing

Many shepherds have come by years of experience to know, perhaps without being able to explain why, when a sheep is in good condition. Like most things in life, this can be taught, and I have been on two ADAS half-day courses where pens of sheep labelled between 1½ and 4 were available for handling, and then another twenty or so were available for a little examination. Under those conditions, it is surprising how close you can get to an agreed number for the grade based on the following principle:

The area you are assessing is the transverse processes of the loin only, as it is now appreciated that the condition of the dock is irrelevant, and the spinous processes (the backbone) are much less relevant. The hand is placed with the heel of the palm on the spinous processes, and the fingers wrapped around the loin. It is good to try to use the same hand—if grading two sheep at a time with both hands, the non-dominant is often more generous than the dominant hand!

Grade 0 On the point of death, skin and bone.
Grade 1 Transverse processes are sharp. Fingertips can easily pass under them and lie between them.
Grade 2 Spinous processes are smooth and rounded, with only a little elevation felt as corrugations.
Grade 3 Individual spinous process bones felt only with pressure.
Grade 4 Transverse processes cannot be felt. Even the spinous process is a mere hard line.
Grade 5 There is now a depression over the spinous processes.

In practice it has become customary to grade in half-grades, and in a group of twenty or so flockmasters of varying experience, and an ADAS sheep expert, it was rare that disagreement hinged on more than half a grade.

As with all of life, the more you practise, the better you get.

Lambs There is no doubt that in order to show lambs competitively in the summer shows, they have to be born as early as possible. However well you feed a late lamb, it will always be a great deal smaller than the early stock. People who do best in Southdowns do so with lambs born in January and then extremely well fed. Supplementary feeding will have been started at around ten days of age and pushed to the limit, including cereals and molassed sugar-beet pulp. I once heard a wonderful conversation at the Rare Breeds Survival Trust Show and Sale, which went something like:

> *Novice to experienced farmer:* 'Your lambs are so much bigger than mine, I think I need to get a bigger ram.'
>
> *Experienced farmer to novice:* 'What you need is a bigger bucket!'

This is not entirely true, however, as lambs can, of course, be overfed. They should only be fed to appetite, little and often rather than large meals. There should always be plenty of roughage and water, since it has been said that a lamb fed on a high amount of cereal can suffer from bloat, scouring and, finally, kidney stones. Some people get round this by asking the feed manufacturers to increase the salt content of the feed, which makes the lamb drink more.

My very limited experience has led me to believe that feeding the mothers well, and thus providing a good supply of milk in the first three to four months of life, has been the most important factor in getting the lambs away quickly. ADAS have demonstrated that for every kilogram of milk solids consumed, a lamb will put on a kilogram of live weight. There is no doubt, however, that if something happens in a lamb's life to arrest its growth, it rarely seems to recover completely, so in my view, it is better to proceed with caution, the aim being to provide only enough protein to be able to produce the bone and muscle that the growing lamb needs. Any excess protein will be turned into fat, which is not the object.

Finally, on the subject of lambs, it would clearly be unkind to take lambs in the process of weaning to a show, so this should be done several weeks before the show date.

Shearlings There seems no doubt that a little bit of supplementary feed given daily throughout the year produces the best results for such as Down breeds, on the basis that it makes sure there is always enough protein for the growth required. For the skinnier breeds, however, a great deal of extra feed seems to be necessary with some sheep needing as much as 3 lb of concentrates a day to put on the sort of condition the show ring demands.

Let's pause a moment to consider what this means. Everyone knows that Down breeds 'run to fat'; as a Southdown enthusiast, I can confirm this. Actually, all sheep can be made fat just as all humans can. It's just that some need more food than others to do so. I can bring my Southdowns up to show condition with grass and a modest trickle of concentrates, whereas the same regime would leave a Kerry looking like a hat-rack. When stuffing in the levels of coarse rations needed for a skinny breed over such a long period of time it becomes even more vital to check with the feed manufacturer about the levels of minerals in the feed. Even though copper is not added to sheep rations it may well be present as a naturally occurring mineral in one or more of the ingredients. Certainly you should also discover whether your soil is rich or poor in copper and adjust your feeding accordingly. I, for example, am on a soil rich in iron ore and therefore I risk swayback, specially in a mild winter. I need to inject my ewes with copper every year and so am unlikely to poison them with small amounts of copper in feed. A friend in Nuneaton, on the other hand, could very easily poison his flock because the copper levels in his soil are much higher. He has found that mixing his coarse ration with a lamb pellet avoids the problem. It has been suggested that this works because the molybdenum in the lamb creep pellet counteracts the copper but there is no proof of this. I am of the belief that it works because the levels of minerals are so low in lamb feeds that the overall mineral intake is by definition reduced. Anyway, whatever the reason, it works. A physicist friend of mine says, 'Theory is often right; practice always is!'

When in any doubt about the wisdom of a feeding regime, why not do as I do and phone the experts. I use my local ADAS sheep specialist, whose opinion I value highly and who, remember, has no axe to grind. Regrettably, local feed merchants rarely know much more than the price and levels of protein, ash, fibre and oil and a suggested ME. Your ADAS expert will perhaps have already taught you to be sceptical about these since a recent survey showed quite a few to be wide of the mark. Speak direct to the actual manufacturers of the feed and you will get a great deal of help. They are naturally bound to be enthusiasts about their own products and keen to sell but, none the less, they do know their facts about ruminant nutrition. They will be happy to discuss feeding with you, particularly if you show that you have done a little homework too. I know that I can get good sense out of Roger Brand of Odams feeds and when it comes to the highly specialised cooked feeds so often used by those in the highest levels of showing, there are few men more clued-up than Martin Dancy, the managing director of TMA Feeds. Trident, the part of the British Sugar Corporation that produces molassed sugar-beet shreds and pellets, will send you helpful literature to enable you to understand where beet pulp fits into the diet. I have found that no feed company looks down its nose at the smaller customer and all go out of their way to be helpful and deliver even modest quantities almost anywhere. I have felt that my custom was valued even when I only ran two ewes.

Of course, as I have suggested before, you can show sheep that have been sheared at the usual time in the late spring, and there is nothing wrong with doing this. Indeed, I am bound to say that I am tempted to try to start a move to make this more generally acceptable. I feel that not only is January shearing rather brutal, even when sheep are housed, rugged and well fed, but to have to carry a full fleece through the summer months must also be very unpleasant for a sheep. While sheep often pick up in condition once shorn in January, one would imagine a comparable loss of condition in late summer. Perhaps carrying a full coat in summer is why it is difficult to maintain a sheep in show condition for any length of time. The aim seems to be to work towards a peak of condition to coincide with the show.

One shearling ewe that was sold at a Rare Breeds Survival Trust Sale in September, beautifully trimmed in heavy wool, had to be sheared by her proud new owner because she became frequently cast!

From the in-lamb ewe we know that cereals must be introduced very gradually and built up to allow the rumen a chance to accommodate. What we don't want is bloating, choking or scouring, and on a purely financial level we don't want to load so much cereal into the system that it merely falls out of the other end. Along these lines I heard a revolting story about a farmer who fed so much corn to his hunters, that the pigs would follow behind and eat the dung, it was so rich in nutrients!

Adding molassed sugar-beet allows for a greater rumen usage of any feed, but particularly the hay and straw and so reduces the need for the high protein to some extent. It is said by some authorities that it takes six weeks for the system to adjust fully to a cereal diet, and it certainly takes six weeks to raise a dry ewe at flushing by one grade level.

Here, a full understanding of the grading of sheep is vital, because at all times along the path to the peak of condition you should be confident to frequently place a hand on your stock and know whether they are at the right stage, and whether you are improving the condition at the right speed to achieve the object.

As a rule of thumb, you might aim for about a kilo of feed over and above forage (hay or straw) for ewes up to 60 kg at full size, and at least 50 per cent more for rams. Rams are often more active in the pen than ewes, so need more feed and take it well. Ewes eat more hay after shearing, and seem to eat more at shows.

With sheep, every change must be fairly gradual, to avoid upsetting the rumen. On returning home from the show, do not just turn the sheep out onto grass, but harden them off over a few days, reducing their cereal and perhaps allowing them out to graze for half an hour or so a day.

ADAS suggest that a 60 kg shearling ewe not in lamb has the following dietary requirement to effect the necessary 90 gram weight increase per day:

Metabolisable energy 12 MJ/day
Crude protein 150 g/day (assumed digestability factor of 0.7)
Minimum forage D value of 57°
Minimum dietary energy level M/D 8.5

The metabolisable energy value of a feed is that part of its total energy which can be utilised by an animal for maintenance and productive purposes. It is expressed as megajoules of metabolisable energy per kilogram of dry matter (MJ/kg DM). D value is the digestible organic matter content of feed expressed as a percentage of the dry matter. M/D indicates the metabolisable concentration of a whole diet, and can be applied to a single feed, or more usually to a mixture of several feeds. It is expressed as MJ/kg DM. With a single feed M/D = ME of the feed (metabolisable energy).

Let us put this into perspective when we analyse the actual feeds available to us in terms of this value:

	ME (0.1150)	CP	Dry Matter %
Good-quality hay	9.3	11	85
Poor-quality hay	7.5	8 or less	85
A typical 18% compound feed	12.5	18%	87
Silage	10.0	9	20
Molassed sugar-beet	12.5	11	90
Grass, spring	11.2	6–8	20
Straw, barley	6.7	1.5	86

Clearly these feeds vary considerably in cost:

	Cost/unit energy in pence (1987)
Straw	0.69
Hay	0.81
Silage	0.83
Compound feed ⎫ Sugar-beet ⎭	1.29

These figures were valid in 1987, and various factors, inevitable in agriculture, will alter them slightly over the years. The soya crisis and the difficulties surrounding the banning of the addition of animal residue products in 1988 will have affected the prices of compound feeds, while the drought of 1989 will have raised the value of hay and barley straw. These factors, however, won't be so extreme as to spoil the obvious point that forages are the cheapest way to supply energy, but less protein, so if a housed shearling ewe needs 150 g protein a day, she may be fed 0.9 kg 16 per cent cereal nut per day, which would be expensive. One kilo of hay, costing much less, will provide her with 110 g protein and rather less energy. Besides, she would be unlikely to want to eat as much bulk.

The rule of thumb is that a non-pregnant, growing ewe needs 2.7 per cent of her own weight per day in dry matter (in late pregnancy this reduces to 2.1 per cent, but rises to 3 per cent for lactation).

A 60 kg ewe therefore needs 1.62 kg DM daily. If she only has hay, this will produce 129 g protein, and all her energy requirement, but clearly falls short of the correct diet.

The answer is a mixture of both the low-value, high-bulk forage and the easily assimilated cereal-based feeds. Adding 15−25 per cent molassed sugar-beet does not reduce the cost, but makes the feed more succulent, more interesting, and as said before, sweetens the rumen.

Finally, sheep are very sensitive to stress, and in the same way that you starve them before dipping and shearing, you ought to lock them up without food overnight before carting them off to the show. Once settled at the showground, a modest feed is appropriate, but remember not to feed too much, particularly cereals. Many people only feed hay with perhaps some succulent such as sugar-beet or rape or turnips, depending on what they have been used to. No one wants to see a bloated, sluggish, overfat sheep at a show, but some level of peaceful contentment coupled with an alert keen appetite and manner is about right.

The Beauty Parlour

Some breeds of sheep lend themselves to specialised preparation for shows. There was a time when this area was full of closely guarded trade secrets, but nowadays those that believe there are any real secrets are deluding themselves. No one extra ingredient could possibly improve the look of a sheep that much. A good sheep, well prepared, clean and healthily turned out will always impress the judge. I have even known people to win championships with dirty sheep apparently straight off the field. One man took every prize going at Leicester with a dirty ewe with lambs at foot running round the ring, while I stood there with an animal I had performed the complete works on and came nowhere!

This, however, is not quite the point, which is that there is a great thrill over and above any consideration of winning ribbons to be had by working on your sheep to make it look 'special', adding a touch of glamour, if you like, to an otherwise good animal.

There is, to my mind, very little element of 'faking' about preparing a sheep. Surely, no judge can be fooled. If you can deceive his eye, and I doubt it, you cannot deceive his hand. The judge spends a great deal of time making up his mind, partly because it is quite a difficult task, and also because he knows that *he* is under scrutiny both by the watching public around the ring and, more importantly perhaps, by the handlers in the ring. He knows that his decision matters a great deal, and he does not take this responsibility lightly. An experienced cattle judge I know says that frequently, when he first enters a ring, his eye is taken by one beast, and he then spends the next thirty to forty minutes trying to convince himself why he should *not* favour that particular animal!

A well-turned-out sheep is, in a way, more a compliment to a judge than a confidence trick, and the judge will repay the compliment by taking the time to evaluate you carefully and thoughtfully. It is, after all, from the ranks of those that show that judges are drawn. They know all too well what it feels like

30

to be in your shoes, and they have a very clear view of the level of conscientiousness to be expected when under the eye of a judge.

SHOW DIPS

As you will by now have gathered, the various different breeds present their sheep in different ways, presumably in an attempt to demonstrate their specific desirable qualities better, though in some cases it does appear to be that tradition plays a big part. Take the colour of the wool, for example. Most non-coloured wools wash out snowy white with Lux flakes, but you will have noticed few sheep at shows looking white. Suffolks, for example, look a deep beige, which to my eye is a rather attractive accompaniment to the silky black face and legs. How and why is this achieved?

I suspect that attractiveness has little or nothing to do with it. It is important to Suffolk breeders to demonstrate the meaty conformation of the carcase and that, even more in this age of man-made fibres, the wool is of very secondary importance. To this end, therefore, the wool must be kept very full-bodied to accentuate the underlying shape. The Southdown is famous for its fine white wool, and so rather more effort is made to demonstrate this feature of the breed. This is my theory anyway, but I should be most interested to hear other ideas. The truth may be that the reasons why certain practices gained favour has been lost in time, but the tradition lingers on, and why not?

Dips are available to both cleanse and firm up the shortwools, and impart the purl to the longwools, quite apart from their usual insecticidal use. Some are stocked by wool factors and agricultural merchants, but others, such as those produced by Youngs, must be obtained direct from the manufacturers.

To produce a darker shade of coat, you would use a phenolic dip such as Youngs Ektamort with perhaps some Jeyes fluid added. For sheep that are shown as white as possible, use a clearer, less phenolic dip such as a scab-approved summer dip with Dettol added. If the show stock are to be dipped statutorily with the rest of the flock, it is a mistake to believe they must go into the dip first. Better results have been achieved by allowing twenty sheep of the main flock to go into the dip first.

There is no need to fill a 200-gallon dip at all, and it would be quite ridiculously expensive to do so just for a few show animals, particularly if you wished to redip them a week or so later to freshen the fleece for another show. In this situation what is needed is the sort of galvanised steel bath you see plasterers mix plaster in. The sheep is cast, lifted into the bath by the legs (a two-man job) and rolled in the dip with the shoulders kept down. Be very sure that the dip is well mixed, and that no sediment has collected on the bottom, particularly if you have used show dip powder.

Show dip powder   This is an additive that puts some body into the wool, useful particularly when you have washed in Lux. It has a cleansing action, but restores the body. While it is pretty expensive, it does at least come in 44-gallon-mix sachets, each one just right for the tin bath, and sufficient to treat eight sheep. This dip would be useful either on its own after a full fleece has had all the bits and pieces carded out of it, or after shampooing to restore some body.

SHAMPOOING

If a sheep has been January shorn, kept in, perhaps rugged and fed a good deal of oily feed, the fleece will be pretty greasy and blackish. In this case, it is a good idea to wash with something to cut the grease about a month before the show, to allow a little natural oil to return. Many would use Lux flakes, believing it to be gentle to the wool, and perhaps more importantly, to the sheep's skin. Others use Fairy Liquid and proprietary animal shampoos. In my view, and this is the doctor in me speaking, all degreasing agents, be they detergent or otherwise, can damage skin and hair in some animals (and I include humans—we're all the same really!). Certainly the biologicals seem to do the most harm, so ought to be avoided. Whatever you use, be very thorough with the rinsing.

Some breeds advocate a wash between ten and fourteen days

prior to shows, so clearly this is to achieve the greatest whiteness and grease content will then be minimal. Some sort of show dip powder might fit in here to restore body to the wool after the shampooing, so long as it does not darken.

This is an area, perhaps above all others, where taking advice within the breed society pays off best. I have spoken with some of the members of several breeds, and the following notes may be helpful. They are not comprehensive, nor holy writ!

Shortwool sheep Preparation begins with a wash around two to three weeks before the show, depending on a really good drying day. Fit a halter and tie up close. Thoroughly soak the fleece, then apply a solution of your chosen cleansing agent, and work up a lather. I find that one sheep, particularly if very greasy, will use at least half a packet of Lux, which needs to be very well mixed in warm water. Be very careful of eyes and ears, and rinse obsessively. I prefer to do this with buckets of warm water, but I don't suppose the cold hosepipe does harm. I just wouldn't like it myself; and besides, warm bucketfuls of water are a more efficient rinse. Finally, I generally give a small feed and turn the sheep out onto a fairly clean field. Do not, on any account, be tempted to put wet sheep into a poorly ventilated shed, or you are asking for trouble. The sheep always shake water off, so you will both be drenched by the end of the washing process. Later, when you check the sheep, you will be appalled to see that they have rubbed against fences, rolled over and seem absolutely filthy. Don't worry, all will be well after trimming, though you alone will ever see the 'white-as-driven-snow' effect that you have just produced, since over the next ten days the fleece will gradually absorb dirt from the atmosphere.

Washing, as I have said, is not absolutely essential, but it does show the whiteness of the wool and, again, I believe it is a compliment to the judge to ask him to handle a clean sheep. You could leave your presentation at this, but it really does make a wonderful finish if you can learn to trim.

Trimming

Properly trimming a sheep occupies three days, so you have to be well in time. To prevent staining at this stage after you have worked on the sheep, you should use a light cotton rug, which you can make out of an old bed sheet.

I love the sight of a well-trimmed sheep, and I am proud of my ability to turn a sheep out well. No written description can possibly teach trimming, but I can pass on a few points that I have learned. I was taught by the master, John Randall, but there must be several other people who can be watched trimming at shows, and usually they are only too happy to show you how they work.

There are, of course, critics of trimming who say that the 'backing down', for instance, of a shortwool sheep is designed to fool the judge into thinking that the sheep is longer and wider than it really is. The reality is that it allows the judge's hand onto the back much more readily. It actually removes the intrusive wool. If there is something wrong with the sheep's back, the judge will see it much more quickly. You therefore tend to take a great deal more off when you are preparing for fatstock shows like Smithfield, than you would for, say, the Royal Show.

When you are trimming, of course, it is essential that the sheep stands still. This can be difficult to achieve, but is much easier if you use a trimming stand, which is an adjustable face restraint with a chain that passes behind the head. The sheep seems to accept this much more readily than just a halter, and will stand still for hours at a time, particularly if sympathetically handled. The fleece must be well wetted all over with a water brush, and then thoroughly carded with a well-rounded wool card with pointed teeth, not flat like a spinner's card. The carding action teases out all the little twists, and produces a sort of 'afro frizz'! Carding is the real secret of trimming, and spending at least twenty minutes to card initially is well worth the effort. Do use a water brush rather than a spray to wet the wool, because a spray just sits on the surface, whereas the water brush puts the water a certain depth into the wool and makes the use of the shears a great deal easier.

The trimming can be divided into two main sections: the actual 'trimming' producing the shapes, effectively rendering the back horizontal and the front and sides vertical, with no attempt to round or follow the contours of the body, and then the 'facing' creating the final felted overall appearance.

TRIMMING EQUIPMENT

As with almost any kind of practical endeavour, having the right equipment for the job and, of course, keeping it in working order, goes a long way towards a successful outcome. The following items are the absolutely basic requirements:

Halter Much of the work is best done with the sheep held in a halter, its chin resting across the left forearm (if you are right-handed). Most particularly, I find the squaring of the beast easiest this way, and I know some prefer to back down (flatten the back) using this hold. I am much less competent at backing down, and so definitely need a trimming stand.

Trimming stand (*see diagram*) This invaluable implement holds the sheep still wonderfully well, and is a real 'second pair of hands'. There are several variations, mostly relating to the way they are adjustable for height. You will see people using them at shows, mainly just before the judging, but occasionally later if they have, because of pressure of time, brought a sheep to the show that is destined to appear at a later date at another show. Decide on the type of stand that seems best to you, and find out from the owner where they obtained it. You can't buy them in shops, but several exhibitors act as agents for the various blacksmiths that make them. I got mine through John Randall of my breed society, who I know generally takes a few to the Royal Show in case there is a demand. Wally Powner sells a really splendid stand complete with a raised platform, hinged at the front end so that it acts as its own ramp, up which

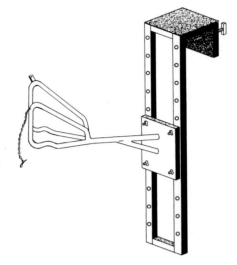

Trimming stand or, perhaps more accurately, a headholder

the sheep walks. Once the head is secured in the restraint, the platform is restored to the horizontal, leaving the sheep comfortably secure at a height that makes the task a great deal less tiring. Naturally, such a sophisticated piece of kit costs a good deal more than a simple stand, but if you were to be heavily involved as some folk are, trimming dozens in a season, it would save a great deal of time and backache.

Shears When I first watched trimming, I was terribly impressed by the ease with which the shears were used. Click, click, click they went, tirelessly across the fleece, for long periods of time. Indeed, all over the sheep sheds you would hear a veritable chorus of clicking. When I tried I found them stiff and cumbersome, and came to believe that I should need to strengthen my fingers and wrists to perform as the experts did. It was only after two years struggling that I twigged everyone else was using double-bow shears and I was using single-bow! The difference in power needed to close the two types is very noticeable indeed. Now, even I can click away for hours, admittedly producing lovely sore blisters and callouses on my fingers, but at least I am now able to trim effectively.

Some double-bow shears are sold with wing nuts to hold

The correct way to sharpen shears

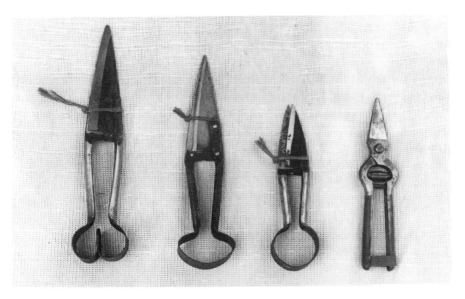

Some tools for the job. Left to right: double-bow shears, single-bow for the masochistic, short dagging shears (useful around the face, ears, horns and tail) and foot-paring shears

the two blades at the bow. These are called regulators, and I presume that they can be released to aid sharpening and setting. Good quality and maintenance is what counts. It is quite amazing how many people seem to believe that tools will serve you forever, even if they are of poor quality and badly looked after. Shears blunt rapidly, and must be sharpened before trimming, unless you want a poor finish. When in doubt, check and resharpen. Always keep a small pot of oil with your shears, to be smeared over the dried blades after every use. They can be wrapped closed in an oily rag, though you often see them stuffed into an old rubber teat.

Sharpening Particularly just prior to backing down, sharpen, sharpen, sharpen. Never fear, you won't wear them away in your lifetime! Take care not to widen the spring of the handle when separating the blades or the shears are lost. Gently, flip the blade edges over, then *overclose* them to expose the bevelled edge to which you are to apply the stone.

The stone is held across the blade edge at an angle of not more than 45°. Do not attempt to produce a finely tapered blade edge, remembering that the two blades act together, not alone like a pen knife. Never, never touch the hollow back face of the blade. Any burrs you have made by sharpening will come off when you carefully replace the blades in their working position, and open and close them a few times. Again, do not stretch the spring.

Comb It is important that this be curved convex, and that the thousands of tiny bent steel teeth are *pointed*. Notice that this makes a trimmer's comb totally different from the sort of comb that spinners use in pairs, which are flat and bear blunt teeth.

A trimmer's comb should be solid and heavily made to stand strain, the block being of elm and the handle of ash. The carding wire is bought separately and secured to wood with tacks after greasing the leather back

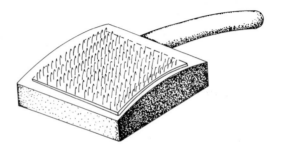

You can muddle through with the spinner's combs but the proper tool does the better job, as always. Again, these are difficult to find, but you will discover that there are several show people willing to provide you a square of leather-backed carding wire at modest cost. I acquired mine from John Randall, and made up the wooden handle and backboard quite easily one weekend out of some pieces of hardwood scrap (*see diagram*). The carding wire square was attached using an upholsterer's staple gun.

Comb cleaner Combs soon matt up and must be frequently kept clean or they soon fail to operate to the full depth. Many people use a bent dinner fork for this job, but by far the most effective and elegant tool is an old shearing machine blade, if you are lucky enough to be able to scrounge one. Failing that, pet shops sell a very useful carder actually designed for use with dogs and some people think this to be the best implement of all.

Water brush This is needed to dampen the wool before carding. The water is then carded in to the correct depth to enable the shears to pass more readily and evenly through the wool at the depth you decide. Do not be tempted to use a spray which tends to both over wet and dampen too variably. The old saying goes, 'If the day is so damp you wouldn't make hay, you shouldn't trim sheep.' What is required, then, is controlled dampening, action with the shears confident and purposeful followed by rapid full drying.

Bucket This contains the damping agent. Water alone will do, but the ideal mixture seems to be half a gallon of water, a handful of show dip powder and either one teaspoonful of phenolic dip or Jeyes fluid for the darker finish, or one teaspoonful of Dettol for the lighter. Remember, wool is a biodegradable material, which only tolerates wetting in nature (or even in a Guernsey sweater) when oiled. The Dettol or the Jeyes fluid may prevent the bugs from doing what they normally like to do to wet wool.

Rug Naturally, after you have spent so much time and effort making your sheep clean and neat, the last thing you want is for the fleece to be soiled, so it makes sense to put a light rug on. You will see all manner of materials used from sacking, old bed-sheets, strong calico to smart, cool and what's more important, wipable woven polypropylene. I have improvised them out of old curtains, but while you need them to be substantial and not too easily ripped, you do not want the sheep to sweat up, bearing in mind that in most breeds you will be asking the sheep to carry more than usual wool in mid-summer.

Patience   I know it sound corny, but you just can't expect any real enthusiasm for all this from the sheep. There's no profit in it for them, and it amazes me that they are as tolerant as they are. When you watch a master craftsman like John Randall trim a lamb, you have to marvel not just at the deftness of his hands, but the way his manner with the animal seems to calm, almost mesmerise, it into a state of docile compliance.

Animals are known to be able to sense our fear and anxiety, and to respond badly to it, usually adding to the chaos and then we can blame them when things don't go our way. Relax, talk to the sheep as you work, be methodical to avoid doing the same task twice and prolonging the process needlessly, have respect for the animal, enjoy the work and don't get too wrapped up initially in the winning or losing side of it all. Even if you make a complete pig's ear of the trimming, at least you've tried, and if the sheep is good, the judge will see that it is good.

THE TRIMMING PROCEDURE

Quite the most important thing to recognise here is that the shears must be held either horizontally or vertically, and advanced with the lower blade moving forwards about ¼ inch at a blow, and being relatively stationary, the upper blade closing on to the lower blade. This is extremely difficult to achieve, and it is worthwhile spending some time with a pair of shears, practising to get a little bit of wrist strength.

Trimming has to be considered methodically. When I first began I was really only facing up, and it took me many hours snipping a little here and a little there, with no clear pattern in

(text continues on page 44)

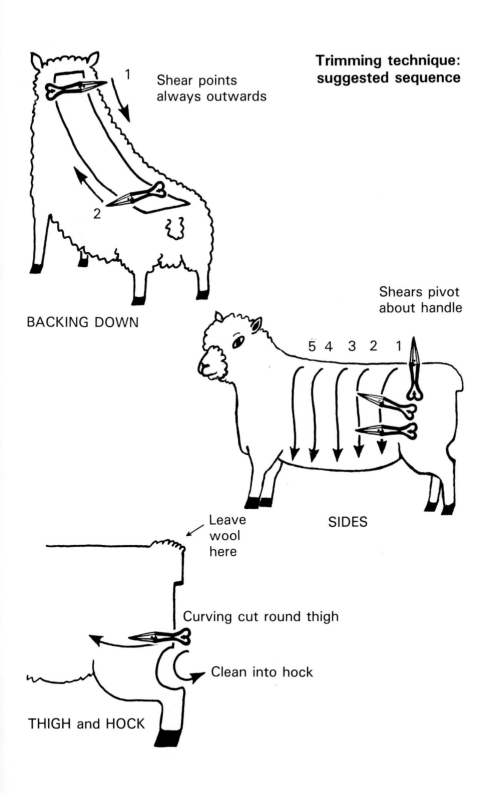

Trimming technique: suggested sequence

1 Shear points always outwards

BACKING DOWN

Shears pivot about handle

5 4 3 2 1

SIDES

Leave wool here

Curving cut round thigh

Clean into hock

THIGH and HOCK

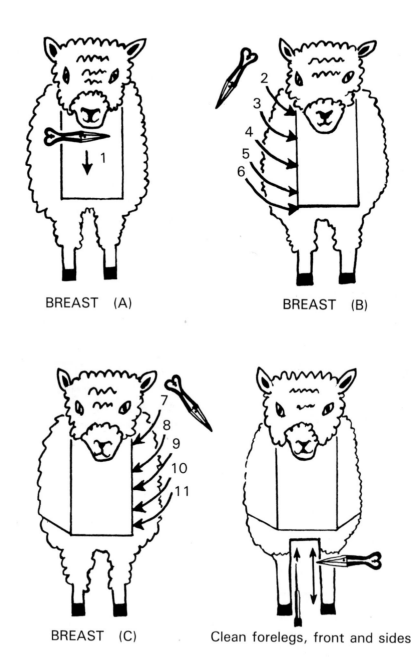

BREAST (A)

BREAST (B)

BREAST (C)

Clean forelegs, front and sides

BACK END

Vertical tail cut (1)

Tail cut (2)
Repeat on
right
side

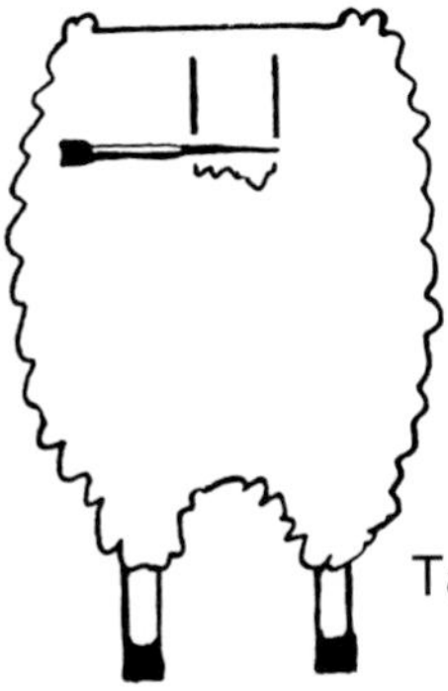

Tail cut (3)

Tail cut (4)
Repeat
on
right
side

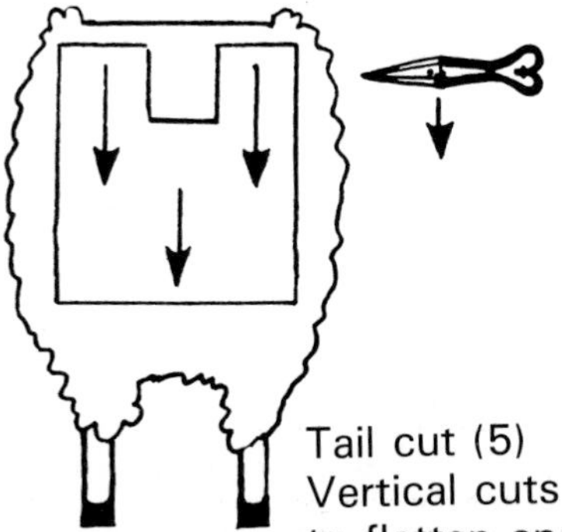

Tail cut (5)
Vertical cuts
to flatten and
square back end

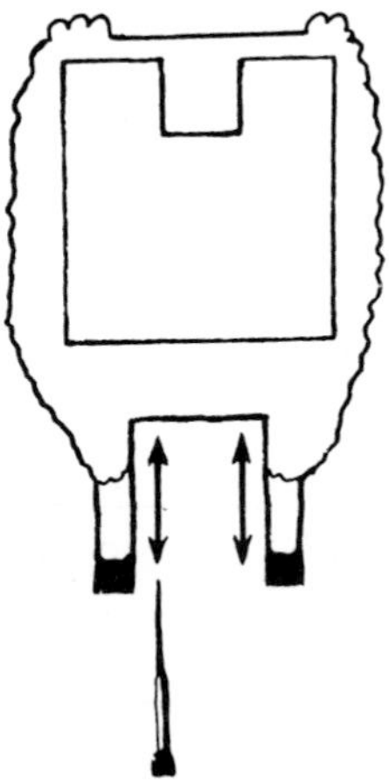

Tail cut (6)
Clean inside
back legs

mind. I was therefore going over several areas more than once unnecessarily. Thinking about the trimming in easily handled pieces, it is convenient to consider the breast, the rear end, the back, the sides and the head, and it seems to go quite well when you taken them in that order (*see diagram*).

Breast   Standing behind the sheep and holding on to the halter with the left hand and lifting the chin up, I square down the breast of the sheep with the shears held horizontally. I do not start too high, in fact I take nothing off under the chin. At this stage the point of the breast is squared off underneath, and the front legs cleaned completely on the insides and at the front, but not the outsides or the back.

Starting on the right cheek, but not taking much wool, I make a semi-circular swathe down and around to meet the right-hand edge of the breast plane. Moving a little backwards, I take a second swathe from just under the right ear, along a similar curve to the same edge of the breast plane, and so on, all the way down to the right shoulder. This then leaves me with a front 'edge' running vertically from the back down to the right shoulder. I repeat the process on the left side, finding it a lot easier this time to stand on the right side of the sheep.

Rear end   I find it most convenient to do the rear end with the sheep held in the trimming stand. I begin with the tail and make a vertical face down the back, square off the bottom and square the sides by placing the edge of the shears right through the wool onto the sheep's rump. The rump is squared vertically, and because of the tail this has to be done in two goes, though you must remember to try to make the bottom look like a single unit and not two buttocks. I clean well down into the area behind the knee, and then take a semi-circular cut from behind the knee around the thigh. At this stage it is usual to clean up the back legs, particularly the inside of the legs.

Backing down   Starting just behind the ears, I take a swathe right down the neck and along the back, with the points of the

(*text continues on page 49*)

*Damping down
with the water
brush after carding*

*Begin backing
down at the neck
keeping the shears
horizontal*

*Progress slowly
but boldly down
the back taking
half-inch bites*

*Start back towards the neck keeping the shear points
outwards*

Flatten the tail as vertically as possible and closely at the tip

Deeply cut down the side of the tail. Sharpen shears first!

Cut boldly and vertically beside tail. Note position of handler

Clean deeply under buttock towards hock. This shows all too clearly that the tail has been left too thick and that the vertical cuts in plates 11 and 12 were too far forward and not vertical enough

Bold curving cut around leg

Flattening the breast. Note the handling position. To have the lamb backed up to the hurdle would be more secure

Squaring off the breast

Flattening the sides or joining the front edge to the back

Curving series of cuts around neck. Beware over bending which may result in deeper cuts than you intended and asymmetry

Don't ever touch this area

John Taylor's home-made trimming stand. Once the sheep has walked up the slope and been secured in the headpiece, the back end is elevated

(above) *A simple portable trimming stand that will bolt on to any gate or hurdle*

(top) *Side view of portable trimming stand*

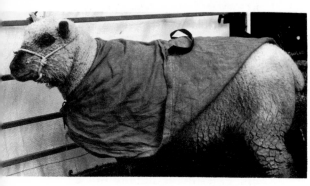

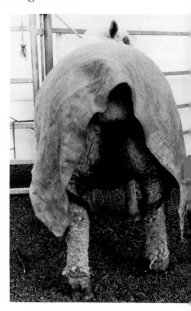

Dolores models a little number run up by the author's wife

(right) *Rear view of ewe coat showing position of tapes*

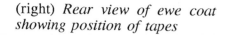

(left) *Commercially availa 'Yowcoat' complete with advertising*

shears facing outwards, generally standing on the left side of the sheep facing backwards. Go no further than mid-hip. I come back the other side, standing at the rear end, again with the points of the shears facing outwards and held as flat as possible. These two cuts are very important indeed, and it does well to take them very slowly, advancing the shears by no more than ¼ inch at each blow. The wool must have been freshly wetted and the shears must be as sharp as possible.

Sides The idea here is to join together by a flat vertical plane the back edge of the front line and the front edge of the back line that you have created by trimming out the breast and the rear end. The shears are held vertically and rotated through 90° to just take off a very light amount all the way along, progressing in 3 or 4 inch sections. I find it easiest to do the left-hand side of the sheep starting from the rear, progressing to the front and attacking the right-hand side in the reverse way.

Always remember to wet down and card the water well in with the carder before using the shears. Finally, dampen down thoroughly again all over, card very thoroughly all over, and then face off the fluffy down produced by this last carding. Try to keep away from those areas that lie between the vertical and horizontal planes, such as the shoulders, just on the hips, and on no account trim off under the 'armpits' or in the groins.

Head Of course, you will appreciate that the whole of this trimming process has been designed to be appropriate for a Down breed of sheep, and finishing off the head is most particularly specific to this breed, though of course there will be similarities across the board. The Southdown is supposed to have a broad head and small ears.

The face should be cleaned in a triangle across the nose, and be of an even 'mouse' colour. The ears should also be of a 'mouse' colour. The wool on the poll, therefore, is flattened, but no attempt should be made to clean away the ears which, of course, would then make them look bigger than they are. The nose is naturally clean, but it is sometimes nice to tidy up the wool edge. I leave plenty of wool under the chin, because it looks better, and I merely face up the sides of the neck. Over

*Mrs Steele, justly proud of her splendid
display of coloured wool skeins, demonstrating
the glory of the Lincoln Longwool*

the following two days the facing process needs to be repeated once each day until almost nothing of the fluff comes on after carding. Between each facing, of course, I rug up to prevent the fleece becoming soiled.

Clearly this process is going to take a good while, and sheep are not famous for their delight at being messed about by human beings. They have to be held lightly, but securely. The tighter you hold them, the more they try to pull away. Because of this, a trimming stand is almost an essential. I have trimmed without one, but the difference is remarkable.

The most important thing with sheep is not to be impatient, and not to lose your temper. After a while they settle down and are extremely quiet for very long periods, particularly if spoken to all the while and reassured. At the end of a long trimming session, I like to reward the sheep with a few nuts and, perhaps while still in the halter, take him for a little run to loosen him up a bit.

Halter Training

With very little training a sheep, male or female, will accept a rope halter and walk like a puppy dog to heel. Some breeds are more easily trained than others, and males seem easier than females. The secret, as with all creatures, is to start young. If they have been properly trained, they seem never to forget, even after a year or so.

Most agricultural suppliers can provide sheep halters, and they are normally on sale in several different forms at the various shows, most particularly the Rare Breed Show and Sale at Stoneleigh. Quite a reasonable halter can be made very quickly out of

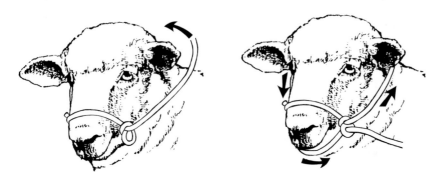

Home-made sheep halter

any piece of soft rope, as in the diagram and photo. I therefore have many of these informal halters which I use frequently, and only use nice clean white shop-bought halters to show in.

It takes a little while to get used to the idea of how the halter is fitted, and most people seem to have them the wrong way round when they start. The basic principle is that the last place the halter should tighten, i.e. the end noose, should be around the muzzle and not around the back of the head. In a sheep with a lot of wool it makes sense to ensure that the halter is closely

51

applied to the back of the head and not just the wool on the neck; otherwise the sheep very quickly learns to back out of the halter.

I train my show lambs to the halter as soon as they are weaned. To start off, I simply let them try it on and perhaps tie them to a hurdle, or just stand and hold several lambs on several halters around me. To begin with they pull and jump and kick, but very soon they realise they are not going anywhere, and come to stand quite still. If you then decide to move them, they start to jump and kick again, so at this juncture I introduce bribery and corruption; using a pocketful of sheep nuts I entice the lamb to come forward to get some nuts from my hand. Very quickly the lamb appreciates that coming forward is good, pulling back is uncomfortable, and eventually they all follow you around very willingly with very little effort on your part. In the weeks leading up to a show, I take my sheep for a daily walk, much to the entertainment of my village neighbours.

Some breeds, particularly the more excitable primitives or the hill breeds, react much more violently to attempts to tame them than my easygoing Southdowns. They take longer to train and pull and leap about more frantically. The outcome will be that a rope halter will chafe, initially rubbing all the face hair and finally cutting a sore area under the chin. The trick that the Kerry breeders use to avoid this problem is to make their training halters from old socks. No, this is not a wind-up—they really do. I would have thought that an old pair of tights would have been more appropriate but I am assured that woollen socks, tied together and attached to a piece of string, are the ideal halter in the circumstance.

The whole process has rarely taken more than a week, and once done a lamb can be readily brought back to full training as a shearling, or even older. Once a sheep has learned to accept a halter, the process of loading and unloading stock into trailers and moving them about the farm individually is so much easier, I am surprised more shepherds with small flocks like myself do not use it.

The same goes for the 'follow-the-bucket' system of shepherding, where sheep are so docile and familiar that I can walk among them and step over them while they are lying down cudding without causing them any anxiety. This is useful in one

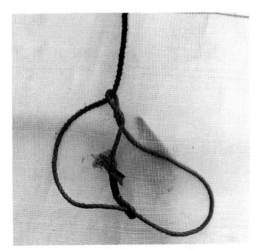

A simple halter in spliced rope. When stuck you can rapidly fashion something similar from binder-twine

respect, but of course means that I cannot chase them anywhere, though I can always entice them by rattling a bucket with a little bit of food. In fact, they are quite happy to follow me down the road from field to field for more than half a mile. This way I do not need a dog and, in fact, manage nearly all my shepherding entirely single-handed.

Show Preparation

I have spent a good deal of time describing how to prepare a Downs breed, and this will hold good for a large number of shortwool types, including Suffolk, Oxford, Hampshire, Ryeland and Shropshire. The following few paragraphs may give a guide as to the preparation of some other breeds, but they are not designed to be comprehensive, and by far the best thing is to join the breed society, and contact the members nearest to you as well as at shows, to acquire the specific information. I have rung around to several members of other breed societies and found them to be most open and helpful.

LONGWOOL SHEEP

I have discussed the principles of preparing longwool sheep with Brian Holgate, the secretary of the Wensleydale Society. I presume much of what he has told me holds true for other longwool breeds, such as Lincoln and Leicester. The first thing he stressed was the condition of the sheep. Although the wool is what you see, remember the judge uses his hands, so much of what I have already said about getting sheep ready will go for long wool as it will for short wool. Wensleydales seem to be shown in two basic ways.

Hogg in wool These are seen at the Great Yorkshire Show during the second week in July. Here the original lamb's wool has been retained for all its life, so it represents about sixteen months of wool growth. You cannot clean the wool, but you do keep it free from straw or soil, if you can. Just before the show you damp the wool down on top, then you centre-part it along the spine and set it to lie nicely.

Shearing after January 1st Apparently this should ideally be done by hand. The belly is clipped out and then the sheep is allowed to stand, beginning at the rear end all but a skirt is

Mr Humphreys of Powys judging pedigree Welsh Mountains

Beatrice, the author's youngest daughter, starting at eight as she means to continue with her own first lamb called Imogen

Mike Pullin's gorgeous Wensleydale ewe. Justly proud, Mike has presented this fine lustre longwool fleece to perfection

A charming pair of Badger-faced Welsh Mountains, or Torddu (meaning black belly). Interestingly, these occasionally appear in a reversed colour pattern when they are called Torwen (white belly)

A strong line of Southdown tup lambs. The author's eldest daughter, Alice (third from the (right), has been showing from the age of eight

Mr Myrfyn Roberts poses a handsome Bluefaced Leicester ram. This breed, like several others, is judged, in part, running free in the ring and is controlled in the line with a hand under the jaw. Halters seem to be hardly used

John Burroughs' North Country Cheviot ram. Bought and brought down from Kelso to work in the west country, he has had plain red dirt worked into the coat to produce a similar finish to his northern competitors

Lonks. Another hill breed, from the Pennines. Are Lonks Derbyshire Gritstones with horns or are Gritstones Lonks without?

Herdwicks 'on tow' showing the difference in tone between those from their normal dull wet Lakeland habitat and those sunbathing in the East Midlands

Serious business, this showing! Swaledales, tough sheep from the Yorkshire moors, produce the Masham and Northern Mule when mated with the Teeswater and Bluefaced Leicester rams respectively

Wiltshire Horn, the breed without wool. Historical records and archeological finds suggest that the sheep of 2,000 years ago were very similar to these

Trimming a Suffolk. Notice the incredibly fine felted finish and the obvious use of a mildly phenolic dip on the middle sheep with a rather open fleece

Cotswolds, a venerable breed which crossed with the Hampshire Down in the 1830s produced the Oxford. Shown here by Libby Henson and Mrs C. Cunningham (right, with the ram)

The highest accolade of all: the Supreme Interbreed Trophy. In the foreground, Supreme judge Mr Beavan confers with the stewards. In the background, Geoff Riby awaits the result with his Oxford ram

Wensleydales both sheared and 'in wool'. In the hat is David Randall, Southdown Champion 1989 and honoured to be asked to judge at the 1990 Royal

clipped out from the back. Two weeks before the show a 'purl' dip is used, and apparently this is an area where folk use very secretive additives, even incorporating substances like peat into the dip. It is important to choose a sunny day for this. The dip tightens the staple and produces an even crinkle along its entire length, which is what the 'purl' is.

A good 'skin' has an even purl all over the body with ringlets like open spirals, as if formed around a central needle. It may be necessary to dip two or three times to produce this effect of tightening the staple, but since it takes three days to dry a Wensleydale, the latest you could re-dip would be three days before the show. At about a fortnight before the show the face is cleaned up, the poll clipped bare and the dock squared across. When the face is cleared the skin will often look pinkish, but there are still two weeks before the show and by then it ought to look a lot more blue.

BORDER LEICESTER

Here, tremendous changes have been seen over the last twenty years, and the breed, traditionally held to be a female-producing type, is tending to become more dual-purpose, with a rather better back end than before. Thoughtful breeding also seems to have overcome the problem of the overshot jaw that was once a strong feature. At the same time, the head has become more alert and attractive, with 'cocky' ears, and it is, with respect to the jaw, more sound.

The aim of preparation for shows is, as ever, to accentuate the selling points. This is achieved by minimal shaping, the accent being on facing-up the wool on as long a staple as possible. This is done dry rather than using the waterbrush technique of the shortwools, and sheep are customarily housed for a month before shows, to ensure this heavy fleece really is dry.

The legs are cleaned, and along with the clean face, are whitened with zinc oxide paste. This practice is harmless and legal in this breed. Compare this with the totally illegal use of blacking on a Suffolk's face, where only a light dressing of clear oil is permitted.

WILTSHIRE HORN

This is the *wool-less* breed of sheep, having a skin similar to goats or cattle. They moult in the spring, so there may be a problem for early shows when, if half-moulted, they look moth eaten and dreadful. The way round this, as with cattle, is to rug them up for a few weeks to accelerate the moult. Do not be tempted to clip!

Preparation is a wash in a cattle shampoo, which does not have to be too rigorously washed out as it is so safe, then brushing to bring out the natural oil onto the hair, to produce a sheen. The horns are oiled.

One point about this breed is worth mentioning. There is a misconception that as a grazing breed they can do well on hardly any feed, and some novices take this too literally. To compete against the best flocks, a good winter feeding programme, as previously discussed, is essential to produce the peak of condition needed for shows.

HERDWICK

The fact that this breed hails from the Lake District and has a long kempy coat suitable for the weather conditions leads to a quite distinctive approach to showing.

The white parts, face and legs, are washed clean to look their whitest, but not whitened with any agent, such as paste. The fleece, however, is reddened using a special red ochre powder only obtained from local merchants in the Lake District, though friendly Lakeland breeders can be persuaded to help breed colleagues elsewhere. The red ochre is suspended in water or, better still, a non-greasy oil, then spread quite thinly over the back from the neck to top of tail, and halfway down the side, shading out gradually so that it does not reach the 'stocking tops'. It should be applied two to three days before the show, and apparently benefits from a drop of rain afterwards to soak it in.

The wool is brushed down to lie nicely but, of course, not

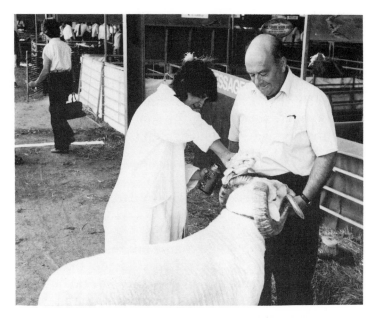

A liberal coating of hoof oil is applied to a Wiltshire's horns

Ben Thomas explaining the finer points of a Wiltshire Horn ewe

combed. Some breeders use a stick rather than a brush. Some dampen, others do not.

The rams are often polled, but if horned, those should be washed and perhaps lightly oiled to enhance their natural creamy colour. Ewes are polled, and must be free from 'slugs'.

The Herdwick is a late lamber when in its natural high environment, so weaning for summer shows can be a problem. One breeder was caught napping by a ewe that lambed at the Royal Show.

BLACK WELSH MOUNTAIN/BADGER-FACED WELSH MOUNTAIN

There is no doubt that if left out in the summer sun these will bleach, so the perfectionists keep them indoors. Others rather like the bleached colour, so don't bother. You pays your money . . . Some wash them—others don't.

They are trimmed lightly, but not squared or backed down at all. The aim is to make the sheep look as if it has just been taken off the mountain, but not dragged through a hedge backwards. I mean neat, but not ostentatiously turned out.

JACOB

These are washed three weeks before showing so that some oil gets back into the fleece. Around a week before the show the fleece is fully carded out. Lambs can be shown just carded, but adult sheep are also usually trimmed. Opinions vary as to the level of trimming, but no backing down is carried out, and often you will see either untrimmed or very lightly faced fleeces. A further trim just before the show day helps to tighten the fleece. Horns are oiled and dried.

KERRY HILL

The outstanding characteristic of this breed is the attractive head with its bold black-and-white markings and fine, cocky

Dick Powell smartens up his beautiful champion Kerry ewe before the supreme championship. Note the chalk whiting on the shears, hands and even trousers! There are different opinions within the Kerry society about the use of whitener. When in doubt, check on the regulations for your breed

ears. For a Kerry to be show-worthy, there must be a good deal of black and in particular the eye patch must be large and oval. Matching up a pair of ewes for showing can be such a headache that breeders occasionally resort to buying one in.

The darker the face the better, but the more likely you are, therefore, to see dark patches of wool on the neck. While these are not actually a fault they are best minimised and there is an interesting and safe method. It has been observed that the patches become lighter at subsequent shearings so that some breeders shear the necks of lambs once or twice leading up to the final show shearing on about Christmas Eve and believe that the density of the discolouration is then less. Others mask the areas by using a chalk white solution on the whole fleece at trimming. While I must admit that a brilliant white Kerry is a magnificent sight, I do side with those Kerry breeders who prefer a more natural approach.

Kerry ears must be fine, upright and on the top of the head. The ideal show position is for the sheep to be looking forward, both ears cocked up or straight back. In handling this is achieved by holding the sheep under the jaw and if the ears droop, slightly jerking the head upwards. To get the ears to lie back, you tickle the wool near the rump. The Kerry is an intelligent-looking sheep and its natural pose when alert is as I have described. After a long show season one early sign that the sheep are 'going over' is when the ears become 'loutish', that is, sloppy and uncoordinated.

Hill breeds retain their tails because there is less need to dock at high altitudes where flies are less of a problem. For show purposes, however, this does not mean that you leave well alone because the tail tips need to be ringed so that in ewes the tail tip falls level with the bottom of the belly. Rams' tails should end at the level of the hock. Tails left this long, as in all hill breeds, are vulnerable to damage, although rarely. It seems likely that ewes occasionally tread on a tail causing it to kink. Thankfully, in showing, these kinks are not considered to be a fault.

Those who breed Kerrys in the lowlands will naturally have to take some pains to avoid blowfly strike and this must be an area where Ciba-Geigy's Vetrazin spray would be a boon. This would also be true for all such hill breeds.

ROMNEY

This longwooled breed is generally shown 'in wool', i.e. with more than a year's wool crop. It is one breed where wool quality is very important, though it is interesting that observers from New Zealand have remarked that some breeders tended to sacrifice wool quality for carcass quality. The point is that the considerable art of preparation for show is to take a good fleece on the hoof and not spoil it! The process cannot be rushed, so that shaping begins as early as March for July. Statutory dipping comes very near to the Royal Show, so that the last step in preparation tends to be dipping. If the sheep is the first in the dip, the outcome is dreadful. The best result is obtained if about twenty sheep are put through the dip before the show stock. Any 'wool conditioner' put into previous dips will be washed out by the statutory dip, and additives to that seem pointless.

TEXEL

The breed society rules are quite specific that Texels must be shown naturally, having been clipped bare on or after May 1st, so that only at shows before this date would you tend to see Texels 'in wool'. Apparently clipping bare means not even using a show comb, though I suspect many people do. No trimming, dressing or shaping is allowed, but you may wash the legs and faces, perhaps with the help of a scrubbing brush. It works well to use Fairy Liquid and wipe dry with a towel rather than rinse, and it makes sense to cover the fleece at the neck with a cloth to prevent dirty suds entering the wool.

In the earlier days the strictures about preparation led to some breeders showing their stock complete with clags, so gradually the society has mellowed a little to allow some cleaning. The sheep is hosed down to soak the dirt well, left for thirty minutes, then hosed again. You might use a dip, again having hose-soaked thirty minutes beforehand. Another thirty minutes and a second dip may also be necessary. The dip bath need only contain water, but if you are dipping anyway, run the show stock through after twenty others have used the dip first.

Here again, because with Texels the accent is on meat, not wool, what counts is the condition and conformation, and less on wool preparation.

DORSET HORN AND POLL DORSET

You will often see these shown in very short wool, and lambs looking as big as other breeds' adults. This is because this breed's outstanding characteristic is that they breed at about any season of the year, producing, if properly managed, three crops in two years, often lambing in November, July and March. This means that the Dorset breeder has great flexibility. The principal show and sale is held in May in Dorchester, in time for the early tupping dates.

The two main categories you will see in summer shows are lambs which, to be competitive, ought to be born in the first week of September, and shearlings, again September-born, sheared on or after March 1st of the following year. By the Royal Show, therefore, they have but four months' wool which, after washing, is carded once and trimmed lightly. The washing is often done in Lux, some use Fairy Liquid or, better still, an animal shampoo. Detergents are generally not a good idea, because they can produce a dermatitis, and take out rather too much of the grease. If you have flattened the back *and* removed all the grease, you had better not put such a sheep back out in the rain, for it will very soon be soaked to the skin. On the other hand, rugging up will cause sweating, with yellowing of the wool! Most breeders house after the sheep is dry, but many allow them to go out so long as the day is fine.

Because soap softens the wool, some showmen dip the clean sheep in something to tighten up the staple and harden the wool. You could use an ordinary dip, but there are show dips, though these can easily impart a yellowish brownish colour to the wool near the skin if not careful. People have experimented with different strengths of clear dip, usually going stronger than normal. Others have used soda in the water, but have noticed the shears blunt rather quickly if they do.

On the subject of shearing, it will be appreciated that if you show a sheep in four months' wool minimally trimmed, it matters a good deal how you shear. The Bowen method, remember, produced long blows along the flanks of the sheep, whereas the Old English or 'left and right' method treats the sheep symmetrically, with the blows following the rib lines. It makes sense after shearing to tidy up with a pair of hand shears, so that a more even growth is achieved. You might even go over again with the clippers (some use barbers' hand clippers for this) a week or so later for evenness sake, particularly on the head around horns.

To return to my theme of showing shorn sheep, a shorn ram class has been introduced in the Dorset sale at Dorchester, because it was thought that trimming was masking quality of rams. Despite the fact that the top Dorset breeders supported the class, it has not been popular, and may perhaps be abandoned.

SUFFOLK

These are backed down before washing, but not too much. Suffolks tend not to be Lux-washed because it softens the wool, but are dipped in a show dip with, admittedly, a packet of Lux occasionally added, though in 200 gallons it can't make much impression on the grease. Again, a really good dry day is important, and if the sheep sweat up, they should be dipped again. In fact, if the same animal is to be shown from say May to July, it may be necessary to dip again before shows. The idea is to tighten the fleece. Those who do use a soap powder often dip or rinse in soda to put some body back into the fleece, so that it holds a good 'face'.

As previously mentioned, the rather attractive beige fleece colour is not 'natural', since Suffolk wool washes out as white as any other Down-derived breed, but is achieved by employing a phenolic dip with perhaps a little Jeyes fluid added, both as a preliminary cleaning dip, and in the damping bucket.

Advanced Showing

Fatstock shows are held in November and December, the most famous of all being Smithfield. What is required is a pair or three matched lambs, depending on whether they are being shown as pure bred or butcher's lambs. Wally Powner has had considerable success both with Suffolks and Suffolk × Southdowns, and because he produces an early mating lamb, he shows lambs born in June. His success relies heavily on the milk produced by the good grass, with perhaps a little creep feeding if he feels the lambs need it. By the end of September he begins to use roots, which he feeds to the lambs for the last two months. He houses the lambs nearer the show, but not too soon, beginning by only bringing them in at night, depending on the weather. He points out that the lambs need the exercise, but on the other hand, since some shows are indoors, they must get used to the new environment without too much stress. The lambs also need to get used to being handled, but not too used—you want them to be reasonably sharp, but not neurotic.

He sorts out a batch of twenty lambs and groups them so that they are treated a little differently. Ewe lambs flesh more quickly than wethers, so they should be kept separately and fed differently. The problem tends to be peaking correctly to achieve the right weight for the weight class. Regular weighing is therefore essential.

The lambs are washed on a good drying day, not too early, but not too near the show. Between two or three weeks before the show works best, but sometimes a month is acceptable. Wally backs down before washing, which he finds gets away from the problem of having two colours of wool, one in the back and one in the less trimmed sides. Wool is, of course, much less important at fatstock shows, since carcass conformation is what the judge is looking at. Besides, there is not much wool on a six-month lamb, so not much needs to be taken off—just a trimming up followed by repeated carding and facing up—as usual, at least three times.

Two final points about preparation that matter a good deal and can easily be overlooked in all the fuss about the wool:

1. Trim the feet, not just from a health point of view, but to ensure that the sheep can stand evenly with a 'leg at each corner'.
2. A week or so before the show, worm the sheep again to prevent the possibility of them scouring on the vital day.

FLEECE COMPETITIONS

Far too few sheep breeders seem to take wool production seriously, according to the Wool Marketing Board, who gripe constantly about how important the industry is and how, with a modest effort, we could produce a better-quality clip. I suppose that the wool cheque is such a small proportion of a sheep farmer's income that he may be forgiven for worrying more about the body under the coat. Those of us in pedigree, however, must concern ourselves about our wool quality, particularly if our chosen breed is important for its wool. It is no good selecting stock for breeding because they have great bodies unless the wool quality is good too and especially if it is 'characteristic of the breed'.

How can we be sure we are getting it right? In the spring we have the wool-on-the-hoof competitions which are helpful and we can watch our own fleeces being graded at the depot if we have a sympathetic and capable wool factor keen to teach. Most importantly though, we can enter our fleeces at shows, some of which act as heats for the coveted Golden Fleece Award, the final of which is held at Smithfield.

Entering these competitions is easy enough and costs are nominal. As always, read the instructions about presentation and time of delivery and re-collection of unsuccessful fleeces. More important than any prize money you may win is the knowledge you will gain from reading the scoring cards with the judge's remarks and comparing your fleece and its scores with other entries.

I tend to select a likely fleece on the hoof and wash it about a month before shearing. I point out to the shearer the importance of the particular animal so that he can shear with just a touch

Winning fleeces with those vital score cards showing just why they were selected

more care. Making sure that the fleece does not become contaminated by dirt or hay and straw, I very carefully collect and wrap it and store it in a sack made from two empty feed bags which will absorb a little. I have seen so many judge's comments to the effect that 'this was an attractive fleece spoiled by presentation or the inclusion of vegetable matter or excessive marking fluid'. These have often come from experienced flockmasters too.

CARCASS COMPETITIONS

Because you need to submit as many as eight lambs all of a similar type and in just the right condition and grading score, this seems to be too daunting a prospect for all but the larger flocks. The classes seem to be dominated by those high-profile breeds whose supporters clearly spend a great deal of money and time keeping us aware of their presence. I know for sure that I can produce carcasses every bit as good as the ones I see in the carcass hall but as yet can't time eight to perfection in July. It occurs to me that within a breed society it should be possible to collect a reasonable 'set' together to at least demonstrate to the world that we can do it. Otherwise anyone who is swayed by carcass competitions might get the impression that either the rest of us are scared to compete or that the only good carcass comes from Continental parentage.

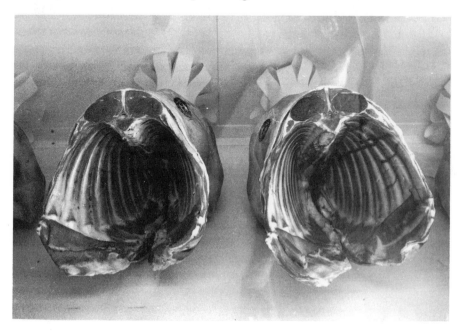

This is what the customer wants: lots of eye muscle and the minimum of fat. Surely, breeding must be towards this end, with showing to demonstrate that we can produce the goods as well as turn out sheep to gladden the eye

SUMMER SHOWS—DIPPING AND FORM M

When you are showing at shows in the summer, some of these, for example the East of England, the Royal Welsh and Kent, will fall in the dipping period. Show organisers ought to send you form M, and you will not be permitted onto the showground without it duly completed. All this is obvious, you say. Yes, but have you actually read the form? What you are declaring is that you have dipped within *twenty-one days*. Suppose you dipped on the first day of the dipping period in 1988, i.e. June 26th, all would be well for the Royal Show in early July, but when you come to enter the East of England showground on July 18th, you are one day over the twenty-one days. Not only is this a nuisance, but you are liable to be prosecuted! Think ahead. Sheep prepared for one show will generally need freshening up for the next a week or so later anyway. They need three days to dry at least.

Mix up a tub with dip at the approved proportion, usually 300 to 1, and with an assistant, 'a leg and a wing', give the sheep a rinse and a second dipping. The fleece will benefit by being tightened up, and you will present them cleaner at the next show, and not fall foul of the law, or be turned away from the showground. You need not be too worried about notifying the Trading Standards Department about the second dipping, because by dipping on day one as usual, you will have already satisfied the scab regulations.

Kit and Caboodle

It is, of course, not necessary to do anything but take clean sheep to the show and trust in the judge's expertise. For me, however, the show is more than just a competition, and so I include a considerable amount of what you might think of as window-dressing. The following is a list of the things that I always take with me to shows.

1. Advertising material I believe this is essential. Unfortunately too few flockmasters seem to want to bother with it. Such material can easily be home-made, and does not therefore have to cost very much. I display a picture which I constructed as a collage from paper glued on to a rectangle of melamine covered chipboard. The picture shows a well-endowed Southdown ram superimposed on a coloured landscape, and the board includes the words 'Dingley Southdowns', made of self-adhesive numberplate letters. In addition to this, I have a name, address and telephone signboard, which was made for me professionally and cost only a few pounds. I also like to display an explanation of the breed history, and incorporate a few of the selling points on a third board. The boards are hung over the pens, generally using binder twine, which does not look unattractive, and I reckon these have paid for themselves many times over.

While I like to be around the sheep a good deal, I also like to enjoy the show, and indeed I just love the whole thing, from the combine harvesters to the bulls, from the cheapjack stalls to the gymnastic display. And I like to visit friends in other areas. Naturally while I am away prospective customers may arrive and seek some sort of attention. Before I leave, therefore, to go on my trips around the show, I make sure that I have left prominently displayed, often on the lid of my box, some business cards and promotional literature. Tons of it disappear every show, and you never know, it may in time produce a contact.

2. Sheets Galvanised pens are always filthy, and clean sheep readily collect the dirt on arrival. For this reason, and also to protect my stock from disease, I have invested in some 12 ft by 4 ft sheets, which I tie inside the pens. They are made of white woven polypropylene, and stamped with the name 'Dingley'. They were made for me by Attwoolls of Gloucester, and although they were quite expensive, I am now a lot happier about taking sheep to shows, and I have noticed several added advantages.

The sheets appear to focus the attention of the often-jaded public on my animals, and I suspect this is because no other animals can be seen through the bars further along. I believe that the problem at agricultural shows is that many of the people become saturated with images of animals, and if you 'frame' your animal within its own pen, it seems to make it look a great deal more attractive. Certainly, many people will walk along the line of pens, and only stop when they come to mine, even though there have been animals equally as good, if not better, in other pens.

3. Buckets   Each pen needs a water bucket, and I prefer to use white ones because I can quickly see that the water is clean and dung-free. I have been very pleased up until now with used glucose lick buckets, though my two young rams have taken a savage delight in demolishing them recently.

4. White coat   It is customary to wear a clean white coat while showing animals. These are available from army surplus stores, and I have found nylon more serviceable than cotton. I also like to wear a hat in the ring, though this is not nowadays essential. The numbers that you see handlers wearing in the ring are provided by the show organisers, either well in advance by post or, increasingly common, on the day of arrival.

5. Lockable box   I have a large lockable box big enough to sit on and store food, clothing, the odd purchases from around the show, and the leaflets that I seem to collect by the ton. I can store my halters, my white coat, in fact everything that I might need around the sheep. I like to use the pen as my centre of activity, travelling to and from it throughout the day, and often the car is many yards away, and not half as suitable to store things. Jane Paynter has the added refinement of a small sack barrow, on

The Daughters pose in front of their Dad's prize-winning display of promotional and educational material

A most attractive sign obviously painted by a professional signwriter. A good investment for the established showman. Even novices should make some sort of effort

What a charming sign. Neat, simple and in such sympathetic keeping with this lovely breed

Betty Garbutt's display is based on a commercially available made-to-order sign that can incorporate a picture of most breed types. Of pressed plastic, it wipes clean and is hard-wearing

which she trundles her box around to shows, but then she is doing shows once a week throughout the entire season with her Southdowns *and* Dexters, so single-handed I cannot see how she could do without one. The Black and Decker 'Move-it-All' would be ideal.

6. Feed   It goes without saying that during the show, animals need feed, particularly roughage, since their energy requirements are fairly low. I feed hay and a little sugar-beet, with very little cereals mainly as a treat, and because they have been taking cereals, and I do not wish to change their diet too much.

Hay nets may save hay, but if you are not careful, lambs can get caught in them and break limbs. It is much better to waste hay by feeding it on the floor than to lose a lamb. Straw is always provided by the organisers at larger shows. When in doubt, check.

A sheep sofa, occupied by an old gentleman showing some surprise at being caught in such an indelicate position! An invaluable piece of equipment for washing undersides and dealing with feet

7. Food and drink for the handler Snacks, of course, are readily available around the show, but they tend to be rather expensive and not always near to where you are based, so it is a good idea to have a supply tucked away in the box in case you cannot always leave when you feel hungry or thirsty. It also provides you with a little something to offer your neighbours in the way of hospitality, which helps make friends.

8. Combs, shears, trimming stand, water brush I incorporate all the kit that I have used to trim sheep, just in case I need to make last minute titivations, but in fact, in reality, it is probably more useful to soothe the nerves before going into the ring than is strictly needed. Some people trim at the very last moment. I have never found this possible myself.

9. Rugs I find it useful to have each animal lightly rugged up with cotton rugs held by tapes around the neck and around the back legs. They spare a good deal of dirt getting into the wool just before judging, though I think it only kind to take them off immediately after they have been judged. My wife runs them up from old bedsheets, though hessian would be longer-lasting, if hotter to wear (*see diagram*).

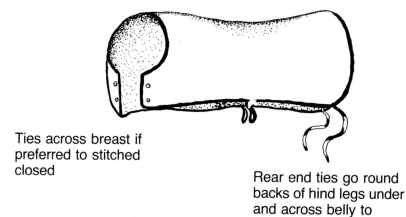

Ties across breast if preferred to stitched closed

Rear end ties go round backs of hind legs under and across belly to loops on opposite side

Home-made show coat

10. Schedule and show programme Show programmes contain a guide time when judging will start on each day, and state which ring certain breeds will be judged in, but since each class may take forty-five minutes or more to be judged, it will be difficult to always know when you will enter the ring yourself. You might feel confident to leave the pen and tour the show, but do not go too far. I have missed the judging myself by doing just this. It is sensible to have the schedule and programme about you, because they will also help you identify the other competitors around you, and help to make friends in the ring.

11. Show halters Smart white halters are available in cotton and nylon, both of which may be washed clean between shows. Nylon, of course, looks whiter for longer.

12. Soft loo paper, soap and a towel 'Enough said'!

Exercise for show sheep If you pen well-fed animals up in small pens for three or four days at a show, they are liable to become jaded and this may show up in their appearance and behaviour. Many of us, therefore, halter them up in the cool of the evening and take them for a brisk constitutional. Others allow them to run free and graze a nibble or two in the empty judging rings. Perhaps it's just sentimentality but it costs nothing and they seem to respond to a change of scenery and the chance to stretch their legs. Indeed, my shearling ram seemed to want to bounce about quite a lot after a frustrating day cramped up in a 6 ft by 4 ft space, and a run did us both good (*see photograph*).

ACCOMMODATION AT SHOWS

As I became more involved in the showing scene, I needed to travel further from home and spend nights away from home when to commute was obviously a nonsense. At first I thought that I would sleep in the trailer, on the straw after the sheep were penned up and, you might expect, after the judicious removal of a few handfuls of currants! Quite a few do this but

Exercise is good for you. The author larking about with his prizewinning ram in the cool of the evening at the 1989 Royal Show

tend to be among the rather younger and more robust among us. A phone call to the show organisers soon provided me with a list of very reasonable bed and breakfast telephone contact numbers, and for shows where I am not accompanied by my family, this has been a superb extra enjoyment to my showing. I have, for example, a regular booking with a lovely lady in Sussex when I visit the South of England Show. I always feel at home, breakfasts are tremendous and I greatly value our friendship. I sleep like a top between crisp clean sheets in a room with a TV and bottled Perrier at hand. So much better than filthy straw!

At the Royal Show, I take the entire family for the week; in fact it's our annual holiday. The kids love it because we stay in a caravan, which is sited alongside all the other livestock owners just outside Gate 5 and within easy walking distance of the stock. There are excellent toilet and shower facilities laid on and you are permitted to bring your own caravan. We find it cheaper to hire for the few times we use one and it saves two trips, one to bring it and one to take it home, as well as the animals in the

One of the many impromptu celebrations and get-togethers to be seen breaking out all over the trailer park after judging

trailer. Again, information is readily available from the show organisation.

Finally, and probably best of all, the serious brigade have the best of both worlds. They convert a cattle lorry or a non-HGV transit van to both carry sheep and double as a comfortable home. Once the stock have been penned up, the straw in the van is cleared away, down goes the carpet and, 'Hey, Presto!' a bijou residence with cooker, chairs and tables. The box over the cab often contains the bed. Up to a certain size these vehicles do not require an HGV licence, although some apparently need a higher level of tax. There is usually a fee to be paid to the show for parking in the approved site but there is a saving on B and B expense or the cost of hiring a caravan.

In the Ring

The steward should have checked that you are present, and made himself known to you. He may then disappear, which is generally because he goes to escort the judge to the ring. You will be called into the ring under your class number, and told where to line up. In the more serious shows, it is customary to stand in numerical order to avoid bias, but in any case you begin in a line, standing at the head of your sheep facing the judge yourself, so that the judge's first view of all the sheep is their rear ends.

Never take your eye off the judge. Look pleasant, but do not leer. Even if the judge is known to you, speak only if spoken to. Some judges seek to put you at your ease by being chatty, so respond briefly and appropriately, but do not try to make an impression. It is the animal that matters. Accordingly, ensure that the animal remains alert and does not nod off. Always make sure the judge can see your animal, and that you are never between the judge and the animal. Do not talk to other competitors, except to be polite. Concentrate on the job in hand, which is to attract the judge to your animal.

Watch how the animal stands. The classical pose is believed to be four-square, head erect with hind legs back and well apart. You will see handlers gently tread on a hind foot which has crept forward, to make the sheep replace the erring foot back. I am not sure this matters a great deal, but being active with the animal and making sure that you do not nod off when the judge seems to be elsewhere is the secret!

After the preliminary look, you will begin to detect whether you have a chance of one of the top placings, because the judge will perhaps glance again at you. He (or she) will then systematically handle each sheep, looking at the teeth and wool, and feeling the scrotum of the rams and udders of the ewes. He will spend a fair amount of time feeling the condition of the back, the back end and the dock, and at this stage you may also be asked to display the teeth to the judge, and should feel confident to do this without unduly stressing the sheep. The judge passes on to the next animal. Do not relax. He will often judge animals in

Wensleydale judge Mr Parkinson takes the trouble to explain his decision to the champion, Mr Duffield, and the reserve champion, Mike Pullin

pairs to make comparison easier, especially in a very mixed class, for example, in an interbreed championship. At this level some judges are prepared to accept as read that the sheep judged to be the best in their class by other judges will not have major defects, such as a jaw problem. Others may not.

In addition to the judging in line, many judges wish to see the animal in motion. Some stand in the centre of the ring and will ask you to parade round in a circle, generally clockwise, at the very beginning, while other competitors are arriving. Here is where halter training really pays off, for nothing flusters a handler like a badly behaved sheep that won't move forward. Some sheep lead ahead well. If yours is not keen, slip in behind another. You will find your fellow competitors only too pleased to chivvy up from behind and keep the line going round. Other judges wait until after they have gone all down the line, then ask you to parade your sheep up and back to watch their gait. If the judge is having difficulty in final placing between two contestants, he may want both of you to walk up and back together. He will be looking for a strong active walk, so do not plod. On the other hand, try not to tow your sheep along either.

Many handlers showing sheep crouch down next to their animals, probably in an attempt to minimise their own body size in relation to the sheep and make it look bigger. There is little doubt in my mind that if you tower six feet tall next to a Southdown ram lamb, you do make him look pretty pathetic by comparison. This is not a sales plug but that same lamb 'on the hook' in the carcass hall of an abattoir is just as big as all the others until you look at the gigot which is bigger!

By now the judge has a pretty good idea of the order he wants to place you in, and he may begin to shuffle the line to help decide the top places. Keep very alert for his directions to indicate wherever he wishes you to move, because he will not wish to attract your attention twice.

Finally, after a brief discussion with the steward, he generally indicates his first choice as clearly as possible by touching the back of the animal, often in a flamboyant manner, and not without a sense of relief, I feel. The other competitors applaud politely—a nice sign of sportsmanship—rosettes are given out, and often the judge will wish to chat to the winners in the ring for a short while as the less favoured are led out.

Classes are usually arranged as follows:

Ram—usually shearling, but older rams may be allowed
A pair of ewes—usually shearlings
A single ram lamb
A pair of ewe lambs

Sometimes a pair of shearling rams are shown in addition to the single ram. The winner from each class then goes forward to the championship. This means that, if you have won a class, you must then hold yourself in readiness to re-enter the ring after the judging of the four or five original classes has been completed.

The last individual class winner usually stays in the ring after being judged, rather than leave and return. The judge then selects the best of these winners to produce the breed champion. This is the stage where sometimes confusion reigns and the novice gets caught out, because the judging of the breed champion can

Mike Pullin deftly tips up his Wensleydale ewe to display the wool underneath

be either very brief or take quite a long while if the decision is difficult.

After the breed champion has been selected, the judge looks for a reserve champion. Novices may fail to realise what this means. Suppose the breed champion has been the ram, then he should leave the ring, or at least withdraw to one side, and be replaced in the line with the ram that was placed second in the ram class. Now there is once more a full complement to judge, and the process starts all over again. While it is rare for both breed champion and reserve to be drawn from the same class, it is by no means that uncommon, so remember the golden rule — all animals under the judge's eye are potential winners.

Remember too that in some shows the breed champion will then go forward to an interbreed championship play-off between all the breed champions of all the other breeds. Schedules usually make it quite clear that this will happen, but because it sometimes takes place on another day to the other sheep judging, you may have to look further in the schedule to see the details. In any case, check with the steward about last-minute organisation.

Chairman of the National Sheep Association from 1979—83 and Chairman of Welsh Halfbreds since 1971, George Hughes is a much-respected Southdown breeder and judge, and he has been unstintingly generous with his encouragement

Hugh Clark carefully rations out the evening feed, pre-soaked molassed sugar-beet shreds, a good staple feed for show stock

The *Showman Shepherd: the late Harold James, pictured here with an Oxford, his speciality before retirement. Fondly remembered today as a canny judge and full of helpful advice for the younger generation*

Mr Bowles, daughter Liz and a Shropshire ram lamb with a trimming stand based on the traditional forked stick stuck in the ground. You get backache carrying this one!

Harold Nobes takes time to advise a fellow competitor, Mrs Parkes, about the health of her Lincoln ram, off his feed after a long, scorching day in the ring

The judging process is fun and can be exciting, particularly if you win, but never forget that the show-ring and the sheep lines are a place of learning. It is customary for the judge to be taken around the pens after judging is over, to have another look at the stock and meet the handlers informally. Of course, you will feel fairly shy to start off with, but after a few years of showing most judges will be familiar to you and indeed, in the fullness of time you yourself may eventually become a judge.

It is generally considered acceptable to enquire of the judge how he saw the competition and, more importantly if you have been lowly placed, how you might try to do better next time. Most judges are sufficiently confident to be prepared to justify their decisions and offer you some insight into the workings of a judge's mind. Without this feedback you will, to a large extent, be working in the dark. Were your sheep too small, too fat, too thin? Do they have a major conformation fault, for which the answer is a complete rethink of your breeding policy, and perhaps some judicious purchases of better blood? Do not be afraid to ask, because this information is absolutely vital to you if you are to succeed.

I am pleased to say that I have never been disappointed when

I have sought critical advice of my stock, and with my breed society I have been impressed by the kindness and enthusiasm shown to me by our accredited judges.

I believe we should at all times handle our sheep subtly, quietly, maintaining a sense of dignity after judging, returning the animals to their pen, ensuring there is food and water before going off to celebrate or commiserate. In my experience, this is the time friendships are best struck between competitors, and yet it is often the case that after all the excitement of the judging ring, everybody disappears, neglecting their responsibilities to their animals, sometimes to their neighbours, and almost always to the general public. Unless you have won, there is, of course, bound to be some sense of anticlimax, but one should anticipate this and guard against it. Most importantly, I believe it to be unforgivable, when one has been unsuccessful at shows, to make an early departure before the specified time and in front of members of the public who have paid money to see the animals as much as any other exhibit.

Breeding

Before people get the wrong idea altogether, may I make it quite clear that I do not propose to suggest that your breeding policies are dictated to by showing. You do not breed to show. We only have to think of some of the unfortunate breathing problems in dogs bred for their showability to realise that this is a thoroughly bad idea. We try to breed our best sheep whether for commercial reasons or aesthetic, and then we show our best results. In reality, it may, of course, amount to the same thing.

First, I would like to address those who have no sheep yet and who, as part of their hobby (with perhaps a rare or minority breed), know they will wish to show, even if only to sell well at the Stoneleigh Rare Breed Show and Sale. My advice is to go for the breed and type within the breed that pleases you most. All your subsequent efforts can therefore be directed towards producing more of what you happen to like. Do your homework and look at several different flocks, whose addresses you will be able to obtain from flock secretaries. You will be able to see several different flocks represented at different agricultural shows and at the Show and Sale. If you are a member of a breed society, there will be annual open days where you will actually be able to inspect people's flocks all over the country. Try to like the good ones! At competitive shows you will soon see who has the most sought after and popular stock, and if you ask flock members tactfully, you may even discover why this is.

According to Stephen Hall, there are often hierarchies in breed societies in which certain elite flocks produce rams that are bought by other slightly less elite but also popular flocks. These 'second division' flocks then sell *their* rams to less prominent flocks, who themselves are less preoccupied with ram producing. This process is seen most dramatically in large flocks such as Suffolks, where in something like 75,000 breeding ewes in the country, there are still only a handful of elite ram breeders who top many (dare I say it?) 'bread-and-butter' ram producers. In these elite flocks 50 per cent of the stock rams used tend to be home-bred. This is because they can run enough lines to enable related rams to be used without showing the ill-effects of in-breeding.

What is inbreeding? Even people who know nothing about sheep can, from what they know about the rules for human beings, deduce that breeding too close is dangerous. The reason is that some hidden or recessive bad genes only come to light when paired with similar genes. It then follows that, while close breeding can ensure that good qualities are passed on, so may bad. It is interesting to see how different breeds differ in this. Close breeding in Southdowns does not seem to be too great a problem, so that the practice of, for instance, 'criss-crossing' with just two rams can be successful for several generations. In the Border Leicester, however, it has been observed that you must breed as far away as possible, or you quickly get numerous difficulties. It may be that the Southdown has been inbred and line-bred for so long that most of the bad characteristics have already been bred out.

If the problems of breeding are disastrous, we soon learn and change our policy. The great danger occurs when the problems are minor, and if accompanied by a sought-after benefit, so that instead of culling the offspring we breed from it, we then perpetuate and even worsen the adverse trait. If, for example, you buy a superb ewe with a wonderful wool but feel she is just too small, so you match her with a lovely big ram with unfortunately poor wool, you may achieve your object of a big ewe with good wool, but you are just as likely to produce a small ewe with poor wool, and you may indeed, if this is the foundation of your stock, produce an entire flock with terrible wool. George Bernard Shaw was approached by a well-known beauty who propositioned him saying, 'With my looks and your brain we should have a wonderful child', to which he replied 'Ah, yes my dear, but what if it went the other way?'

The answer to this problem is to have a really tough culling policy. Buy only the best, almost regardless of the cost, and use only the very best ram, perhaps in your first year by hiring him. Do not believe that you can improve a poor ewe by careful breeding (this is the pathway to tears). If funds are short, buy old or draft ewes that are no longer viable in the top flocks. They have not been kept in the flocks to a ripe age for nothing. Of course, they are not being sold for no reason but, as in the motor trade, people still sell good stock *before* troubles are found, and are likely to try and keep the average age of their flock at round the same level. We all have to sell 'off the top' to make room for our new blood. Ewes are usually sold as sound in tooth and udder, and will not have had a prolapse or other serious obstetric problem. Again, the best bought from the most admired flocks are usually the best bet. An old ewe given preferential treatment in a new small flock may get a new lease of life.

After lambing, try to get the seller of the ewe to come and judge your lambs, and advise you which are fit to keep and why. Most importantly, those that are not deemed fit to keep must go with the mint sauce. It is at this stage that the small flock owners can go wrong. The desire to increase their flock influences them to keep poor stock because poor stock 'is better than no stock'. This is the way to ruin the breed. You will be lucky if you are able to put a third of your lambs back into the flock, and some years you may have to slaughter them all.

I think we all have grandiose ideas of 'improving the breed', despite the fact that it may have taken a hundred years or so for the breed to get to the level that we see it today. Improving is a very long business. Destroying is much quicker, and that is why we do it best. Perhaps we should emulate Jane Paynter, who believes that she does not really 'own' the sheep. She feels more like the relay runner whose most vital job is to pass on the baton without dropping it. I must say I have great sympathy with this view. After all, presumably you have chosen a breed because you like the look of it. If you want something else, why not choose another breed? Southdowns, for example, cannot be 'improved' by making them taller or darkening their face. There are always Oxford or Hampshire Downs or Dorset Downs to go to already.

We must also beware not to become so keen on one type of sheep within a breed that we lose some of the healthy variation of genetic diversity, and end up with so much attention being paid to following a fashionable trend to win prizes and acclaim, that we damage the breed. To this end, the Rare Breed Survival Trust is right, I feel, to change their judges annually, and to insist that judging is by breeders out of all the breed societies, and not just a few top names. Perhaps we should take interest, but not great pride, in the fact that at the 1988 Rare Breed Survival Trust Show and Sale one Shropshire ram had sired all the winners.

What is line breeding? As I understand it, this is using, for example, a ram, subsequently his son and then his grandson on the same ewes. In some flocks the lines are taken from a ram and some, as with dairy farmers, use the female line as more important. Some do a mixture of both.

Making up the flock In the autumn ewes must be selected and then paired with an appropriate ram. How is this done? First, look at the ewes and reject any that are unsound, paying most attention to highly heritable characteristics. The most obvious is the jaw. Any that are undershot or overshot must be sold for meat only. One word of caution is important here. The soft milk teeth of the growing lamb are very susceptible to diet, and by bringing in a lamb to feed on hay and cereals, you may in some cases begin to feel what looks like overshooting of the jaw. If you show the animal in this state, you will no doubt be penalised. If the lamb had no jaw trouble at birth and before you brought him in, then while the judge is correct in rejecting or relegating the animal in the ring, you yourself must not overreact. Put him back in the field, and within a very short time the teeth will firm up in their normal place. It is probably the hay that makes them goofy, and it would be a shame to kill an animal for this reason, thinking he was overshot.

Similarly, the question arises of when should you assess a ram lamb for breeding potential. Some breeders trust the first appearance and weight at birth; others are more cautious and wait to see how well they do, weighing them at three weeks, eight weeks and twenty-one weeks. They would be more impressed by the rapid growth of a small twin lamb to twenty-one weeks than a single getting to the same weight more slowly. After all, why are you choosing a ram? Surely it is to sire a batch of rapidly fleshing offspring who, themselves, may never live beyond sixteen weeks before slaughter.

One very notable flockmaster and judge is prepared to hang

on to rams for a year if necessary before sending them to the abattoir if he feels they are unsuitable, and still gets them graded. Most butchers will handle entire ram lambs quite late, so it does not make too much sense to castrate (or ring) your males the minute you see them. Talk to your local butcher, and see how he feels about accepting late ram lamb carcasses.

Other inherited characteristics include inturning of the eyelids and lax or dropped fetlocks. If ewes have bad feet, this is less of a serious problem, but where you have a choice, be ruthless. Bad wool may be more or less relevant, depending on the breed. Opinions certainly differ as to the relevance of coloured smuts on otherwise white sheep, but clearly the breed characteristics, as laid down by the breed societies, will be an important guide.

In Southdowns, we aim for a sheep with a good length, short neck and a decent 'pair of trousers'; in particular, they should be white-woolled, the fine quality of wool and degree and extent of crimp being vital. They should have a mouse-coloured nose and ears. Here, certain differences are apparent, in that some people go for a very light colour, and others go for a darker colour. Within a reasonable range, all these shades are acceptable. What is clearly not acceptable is if an otherwise polled sheep has horns, even a small degree of 'slug'.

Another characteristic which might influence you in making up the flock is the behaviour of the ewe in previous lambing, and to her offspring. What we are aiming for is one type, both in visual terms and in terms of performance. What I mean is that it is possible to select for easy lambing, 'get up and go', and rapid maturity in conversion of grass into meat at this stage. The ewes are then divided into groups, best ewes being put with the best ram, and ewes with other characteristics that one wishes to alter, placed with a ram who is most likely to alter those characteristics in their offspring. For example, ewes with poor wool be put to a ram with better wool, under average-size ewes with a larger ram. Combining these criteria with the added criterion of whether the rams and ewes are related too closely means that for many the autumn is the most fascinating time of the year.

Not all breeders are as unhappy about inbreeding as others. One lady I know scorns my attempts to avoid it as an 'obsession with the Book of Common Prayer'. She and many others are quite happy to use rams on daughters, mothers, sisters, ad lib. They allege that until you get obvious disasters, all you do is strengthen the line and produce a flock uniformity which they deem a major aim of the breeding policy.

For example, if they have a promising ram lamb, they first, while he is still a lamb, put him on his own mother. If what she throws is good, then they use him on the entire flock next season. The success thus demonstrated at least makes one question one's own basic scruples, and I ask myself whether I adhere too much to my cross-breeding policies for the flock's good.

Coming Home

During the summer show season there are likely to be a whole series of shows with a week or so or only days between them. It will be very tempting, if you are showing the same team, to keep them rugged up indoors still fed on hay and concentrates and this may work well at times. In my experience, however, some animals become jaded and flat. They lose their 'bloom' and may be sluggish and lethargic in the ring. Conversely, rams could be irritable and aggressive. I believe that it is a good idea to let them out to graze for a while so that they get a good dose of 'Dr Green', gentle, natural exercise and a fresh look at the world. They may get a little grubby but rarely so badly that they can't be tarted up before the next show. The areas that get dirtiest—heads and back legs—I hose down carefully without soap since I don't want twin-tone sheep. It makes sense to turn them out onto clean land that is not too lush and for short periods to start with. The little devils make a bee-line for every lichen-encrusted scratching-post and mud-hole but it is worth it, I think.

I smarten my fleeces up with the help of a water brush loaded with a wash containing Youngs show dip powder which is frothy and feels slightly soapy when suspended in water. I believe that it basically only spreads the dirt more evenly and thus removes the marks. Unless you rewash the sheep during the season, it is inevitable that they gradually become greyer. You just have to balance up the equation as to whether the show in question matters that much, what time you have available, if it is drying weather and just how dirty they are. I wash twice: once for the spring shows and once for the Royal.

One very important problem with home-coming show stock is that of fly-strike. In past years I have loaded a knapsack sprayer with dilute dip solution and sprayed the critical areas of head, shoulder and back-end. It did not affect the appearance but the fleeces needed to be well wetted for success. This year I have been very impressed with a new spray-on preparation, Vetrazin from Ciba-Geigy, which is colourless, simple to apply,

dose-related to size and lasts at least eight weeks having insinu-
ated itself throughout the fleece within hours of application.
Although Ciba-Geigy make no claim, it is obviously also fly-
repellent. The beauty of it is that unlike spraying with dip which
is less effective than the actual recommended dipping process,
the as-and-when usage in show stock is precisely the same as
normal usage. This year (1989) has been the worst year for fly
I have known with so many of those muggy days we all fear.
All I can say is that I have had not one case of fly-strike. Even
after the damping-down process of smartening-up, the active
ingredient seems to be working.

Rams One more potential headache for the returning showman is that of reuniting the rams after a short separation. They
have awful memories and seem to need to try and kill each other
or at least re-establish the 'pecking order'. Not only is a ram
with a broken head less appealing, but the blood attracts flies
and you could lose the ram from infection. Last year my No 2
stock ram blew up a terrific fight abscess on the top of his head
so that he looked like a policeman! Despite lancing of the abscess
(and that wasn't funny), he tracked infected fistulae down to his
jaw, his front legs and chest. I let about a pint of pus out of his
chest.

Infection will track under the skin because it is so thick that
abscesses have great difficulty bursting through. For this reason
you need to lance very soon. In this case I was so green that I left
it too late. To complete the tale, the ram developed septicaemia
and needed antibiotics and steroids to save him. He lost all the
wool on his head and it has never regrown on his ears, which
are now like a prize-fighter's. After the infection cleared with
weeks of irrigating the fistulae, he became jaundiced with bright,
thick, orange urine. I drenched him with water and molasses and
injected him with multivitamins for days until it cleared.

Today he is as fit as a flea and is about to go a-courting but
the message is, 'Don't let them fight'. Either lock them up close
in a small space together for at least twenty-four hours with a
straw bale or two thrown in for good measure (and even then
they still 'have a go' when they are let free) or, better still, use
an anti-fight 'hat'. This is a leather device which straps over the

The Ritchey Tagg stand showing a Ken-Shield mask to prevent rams fighting and the 'Rolls-Royce' of trimming stands made by Alec Brown, Past-President of the Texel Society

ram's eyes so that he can feed and even mate but because he can't see forwards, he can't aim. I have used two such devices this year and can vouch for them. It hasn't always seemed necessary to fit them to both rams of a pair because although you might imagine that the 'blind' one would be at the mercy of the other, the reality is that it takes two to tango: the 'blind' one doesn't appreciate that he is in danger, so fails to react appropriately, and the aggressor either can't believe his luck to have met such a coward or is confused and halts his charge. A touch of shoulder barging ensues, thus satisfying honour. You may imagine, after the hassle I experienced last year, how delighted I was to buy two Ken-Shield masks from Ritchey Tagg at the Royal (*see photograph*) and be able to turn out my first prize-winning ram with confidence.

Near to the subject of rams, let's not forget ram lambs. Their fighting is irrelevant but many breeds can be actively fertile very early in life and though they won't work the ewe lambs yet they will almost certainly deal with any ewes that come into season. When transporting the team in a small trailer, I always tie up the ram using a big cattle halter that won't slip or find itself wrapped round the neck of the stock running free. If I believe the ram lambs to be mature enough, they must be tied up too, unless the trailer is large enough to have compartments. Unfortunately, most small trailers are not. I worry about trailing with tethered stock and stop to check their welfare frequently but cannot see any alternative short of expense.

MLC Grading

Once you produce fifteen ram lambs a year, it becomes a viable proposition to pay the modest fee to the Meat and Livestock Commission (currently £180 p.a. up to 50 ewes) for their help in selecting the best potential sires. They use a weight-gain chart and ultrasound scanning of back muscle (£2 per animal) to attempt to be more scientific. Many people are sceptical, but from what I have seen in flocks that have chosen this method, their rate of improvement is quite impressive.

Of course, there is no real need to spend money bringing in the MLC team when you can derive most of the information you seek yourself by regular weighing. The scan option is not available in the small flock scheme anyway, though your local

Surely, here lies the future. Scanning has been invaluable to me to know for sure what number of lambs to expect and feed the ewes appropriately. Back-fat and eye-muscle scanning of potential terminal sires must help us select those most likely to meet the challenge faced by the sheep industry. Show success is not enough

scanner would help here perhaps. Since you will be doing all the actual weighing for MLC in any case, the work is exactly the same. In a small flock the computer is merely a glorified calculator to assess growth rate, and your numbers will not be really adequate for good 'crunching'.

I weigh at birth, three weeks, eight weeks and twenty-one weeks, by which time, with the added use of hand and eye, I am confident to select the best performers. I am surprised how few colleagues bother to do even this little work. Of course, we are all busy men and women, but if you are to worm every three weeks, and probably foot-bath as well in a wet summer, why not weigh at the same time?

The Last Word

In conclusion, may I stress that much, if not all, that I have said here is one man's slant of a wide and fascinating interest. Many readers will take exception to some of the statements made. Well and good—particularly if some discussion is provoked. I hope that I have not misled anyone into believing that between these covers is anything but a taste of what I hope to learn over the next few years.

The message is, I hope, loud and clear. Showing sheep is fun and well worth doing, particularly if one is keen to promote one's own breed and widen the market. Yes, it takes time, but why not warm up to it gradually by showing one ram or a pen of lambs at a small show near home?

So I hope to see a few more beginners have a go, and look forward to meeting you along with my steadily growing family of show friends around the shows.

1989 Show Dates

APRIL	29–1 May	Gloucester County
	30–1 May	Leicester County
MAY	5–6	Notts County
	7	Lambourn
	13	South Suffolk
	17–18	West Midlands, Shrewsbury
	18–20	Devon County
	20	Hadleigh
	21	Horton
	23–26	Royal Ulster
	26–28	Herts
	27	Heathfield
	29	Surrey, Woodhall
	31	Suffolk, Staffordshire County
	31 May–3 June	Royal Bath and West
JUNE	8–10	South of England, Ardingly
		Royal Cornwall
	13–15	Three Counties, Malvern
	16–18	Essex Show, Chelmsford
	17	Derbyshire
	18–21	Royal Highland
	20–21	Cheshire County
	21–22	Lincolnshire
	24	North Yorkshire County
	28–29	Royal Norfolk
JULY	2	Blaston
	3–6	Royal Show, Stoneleigh
	8	Wiltshire County
	9	Midland Counties Rare Breeds
	11–13	Great Yorkshire
	11	South Beds, Toddington
	13–15	Kent County
	15	Newport, Husthwaite
	15–16	Halton, Spilsby
	16	Ashby de la Zouch

97

	16	Royal Isle of Wight, Durham
	18–20	East of England, Peterborough
	22	Cwmdu
	23	Singleton, Mid-west Show, Yeovil
	24–27	Royal Welsh
	25–27	New Forest & Hampshire County
	28–30	Royal Lancs
	29	Abergavenny, Leek
AUGUST	2	North Devon
	3	North Wales, Honiton
	2–3	Bakewell
	5	Brecon
	5–6	Lymme Park
	6	Malmesbury
	9	Airedale
	13	Fillongly
	10–11	United Counties, Carmarthen
	15–16	Anglesey
	15–17	Pembrokeshire
	19	Bedwellty
	23	Merionydd
	24	Monmouth
	26	Barrowford
	27–28	Epworth
	28	Burniston
	31	Monmouthshire
SEPTEMBER	2	Moreton-in-Marsh, Dorchester
	8–9	Taunton
		Rare Breeds Show & Sale, Stoneleigh
	7	Bucks County
	16–17	Hayfield, Newbury
NOVEMBER	21–22	Birmingham Prime Stock Show
DECEMBER	4–7	Smithfield Show

This list does not pretend to be comprehensive nor is it designed to represent a 'Good Shows Guide'. It includes the major events and many smaller ones, gleaned fairly quickly from the pages of *The Sheep Farmer*. Shows are usually held in the same week or weekend of the month even though the dates change each year. For example, the Royal is always from the first Monday of July onwards for four days.

Breed Societies

On the face of it, it would seem appropriate in a book such as this to include a list of breed society secretaries and their addresses. In my research it became apparent that, while many societies have secretaries that seem to have held their posts for many years, such as our own dear Clive Pritchard and Eric Halsall, others change their secretaries as often as most of us change dip-water.

By far the best advice I can give is to contact the National Sheep Association, to whom the breed societies are affiliated, on 0684 892661 or at the address given below under 'Further Reading'.

Further Reading

PERIODICALS

The Sheep Farmer   Published monthly by the National Sheep Association, The Sheep Centre, Malvern, Worcs WR13 6PH

The Ark   Published monthly by the Rare Breeds Survival Trust, 4th Street, NAC, Stoneleigh, Kenilworth, Warwickshire CV8 2LG

BOOKS

British Sheep   An illustrated guide to breeds, incorporating contact names and addresses of breed society secretaries, published by the NSA.

Keeping Sheep   Elizabeth Downing, published by Pelham, 1985

Sheep Management and Production	Derek H. Goodwin, published by Hutchinson, 1987
A System for Lowland Sheep	ADAS Book No. L372
Feeding the In-lamb Lowland Ewe	ADAS Book No. L636
Maedi Visna	ADAS Book No. L659
In-Wintering the Lowland Flock	ADAS Book No. 32065
The Shepherd's Calendar	Val Stephenson, published by Alec Paris Publicity, 1981
Pigs in the Playground and **Calves in the Classroom**	John Terry, published by Farming Press, 1986 & 1987

Sources of Supplies

Attwoolls of Gloucester
Whitminster, Gloucester GL2 7LX   Pen sheets

Alec Brown
Stonefield Farm, Rosewell   Trimming stand
Midlothian EH24 9EB

Ciba-Geigy Agrochemicals   Vetrazin pour-on
Whittlesford, Cambridge CB2 4QT

Alfred Cox Surgical Ltd   Shears, etc.
Edward Road, Coulsdon
Surrey CR3 2XA

Dalton Supplies Ltd   Halters
Nettlebed, Henley-on-Thames   Ear tags
Oxfordshire RG9 5AB

Betty Garbutt Street House Farm, Loftus Saltburn, Cleveland TS13 4UX	Head holders Wool carders Showing courses
Bill Johnson The Stables, Tur Langton Leicestershire	Scanning services
Jackie Kettle 20 Main Street Sewstern, Grantham Lincolnshire NG33 5RF	Sheep show coats made to measure
John Randall 2 Charity Farmhouse, Litton Cheney Nr Dorchester, Dorset DT2 9AP	Carding wire Head holders Trimming courses
Ritchey Tagg Ltd Fearby Road, Masham, Ripon North Yorkshire HG4 4ES	Ken-Sheild masks
Sheepway Products (John Christmas) The Shieling, Stembridge Martock, Somerset TA12 6BP	Hay racks Bucket holders
Sherratt Farm Supplies Wem Shropshire	A full range of sundry equipment for showing
Peter Stone Brookvale, Ringmore Shaldon, Devonshire	Sheep sofa
Trident Feeds PO Box 11, Oundle Road Peterborough PE2 9QX	Molassed sugar-beet
Youngs Animal Health Elliot Street, Glasgow G3 8JT	Show dip powders Phenolic dips
Robin Watson Signs Gallowfields Trading Estate Richmond, North Yorkshire DL10 4ST	Signboards

FARMING PRESS BOOKS

The following are samples from the wide range of agricultural and veterinary books published by Farming Press. For more information or for a free illustrated book list please contact:

Farming Press Books, Wharfedale Road
Ipswich IP1 4LG, United Kingdom
Telephone (0473) 241122

An Introduction to Keeping Sheep   JANE UPTON & DENNIS SODEN

The skills and techniques of caring for sheep for newcomers.

Come Bye! and Away!   VHS COLOUR VIDEO

Glyn Jones demonstrates the basic sheepdog training techniques, focusing on the moment when a young dog is first let off the leash in a field of sheep and learns to obey the four commands.

The Modern Shepherd   DAVE BROWN & SAM MEADOWCROFT

Summarises the technical advances of the 1980s and links them to the traditional shepherd's year.

Sheep Ailments   EDDIE STRAITON

A pictorial guide to all the common sheep ailments. The concise text includes a major section on lambing.

Profitable Sheep Farming   M. McG. COOPER & R. J. THOMAS

Chapters on production, breeds, nutrition, management, store lamb feeding, ewe selection and recording, profitability and sheep ailments.

Intensive Sheep Management   HENRY FELL

An instructive practical account of lowland sheep farming based on the experience of a leading farmer and breeder.

A Way of Life: H. GLYN JONES & BARBARA COLLINS
Sheepdog Training, Handling and Trialling

A complete guide to sheepdog work and trialling, in which Glyn Jones' life is presented as an integral part of his tested and proven methods.

Farming Press Books is part of the Morgan-Grampian Farming Press group which publishes a range of farming magazines: *Arable Farming, Dairy Farmer, Farming News, Pig Farming, What's new in Farming.* For a specimen copy of any of these magazines, please contact Farming Press at the address above.